Key Concepts in Tourism

KEY CONCEPTS IN TOURISM

Jennifer Raga

SOCIETY Publishing

www.societypublishing.com

Key Concepts in Tourism

Jennifer Raga

Society Publishing
224 Shoreacres Road
Burlington, ON L7L 2H2
Canada
www.societypublishing.com
Email: orders@arclereducation.com

ISBN: 978-1-77469-392-6 (Hardcover)

ABOUT THE AUTHOR

Jennifer is a Tourism and Hospitality Management Specialist with more than 15 years of experience in the industry, giving academic training to University Students and Local Communities about Tourism, Hospitality and Events Management with Eviatour C.A.. She holds a BA in Tourism Business Management and a Postgraduate Diploma in Hospitality Management awarded from the University of Birmingham UK.

TABLE OF CONTENTS

LIST OF FIGURES

LIST OF ABBREVIATIONS

AAA	American Automobile Association
ABTA	Association of British Travel Agents
ACE	Association of Conference Executives
ACP	Africa, the Caribbean, and the Pacific
AIDS	acquired immune deficiency syndrome
ANTOR	Association of National Tourist Office Representatives
AONBs	areas of outstanding natural beauty
APEX	advance purchase excursion
APT	advanced passenger train
ATC	area tourism companies
ATK	accessible ton-kilometers
ATMs	automated teller machines
ATOL	Air Travel Organizer's License
BHTS	British home tourism survey
BST	British summertime
BTA	British Tourist Authority
BTR	Bureau of Tourism Research
BTS	British tourism survey
BTS-M	British tourism survey monthly
BTS-Y	British tourism survey yearly
CAA	Civil Aviation Authority
CEEC	Central and Eastern European Countries
CIS	Commonwealth of Independent States
CRS	computer reservation system
DET	domestic escorted tour
DIT	domestic independent tour
DTI	Department of Trade and Industry

DTM	domestic tourism monitor
DVS	day visits survey
EC	European Community
EEC	European Economic Community
EFT	electronic funds transfer
EIA	environmental impact assessment
EMS	environmental management systems
ERP	European Recovery Program
ETB	English Tourist Board
ETC	English Tourism Council
ETC	European Travel Commission
EU	European Union
FCU	fare construction unit
FE	Far East
FET	foreign escorted tour
FIT	foreign freelance tour
FTE	full-time equivalent
GATT	general agreement on tariffs and trade
GB	Great Britain
GDP	gross domestic product
GDS	global distribution system
GHS	general household survey
GIT	group inclusive tour
GMT	Greenwich meantime
GNVQS	general national vocational qualifications
GRS	global reservation system
GSA	general sales agent
GST	Greenwich Standard Time
HA	hospitality assured
HCIMA	Hotel and Catering International Management Association
IATA	International Air Transport Association
ICAO	International Civil Aviation Organization
IDD	international direct dialing
IGO	Inter-Governmental Organization

IHEI	International Hotels Environment Initiative
IHRA	International Hotel and Restaurant Association
IIP	investors in people
INSPASS	immigration and naturalization service passenger accelerated service system
IPS	international passenger survey
IPU	inter-parliamentary union
IT	inclusive tour
IVS	international visitor survey
JICNARS	joint industry committee for national readership surveys
LFS	labor force survey
MC	Master of Ceremonies
MCA	Master of Culinary Arts
MCO	miscellaneous charges order
ME	Middle East
MOD	minor operated departments
MOU	memorandum of understanding
MPM	maximum permitted mileage
MRP	manufacturer's recommended price
NAFTA	North American free trade agreement
NGO	Non-Governmental Organization
NITB	Northern Ireland tourist board
NTA	National Tourism Administration
NTO	National Tourism Organization
NUC	neutral unit of construction
NVQS	National Vocational Qualifications
OEEC	Organization for European Economic Cooperation
OR	operational/operations research
PATA	Pacific Asia Travel Association
PEX	public excursion
PIR	property irregularity report
POS	point of sale
PSA	Passenger Shipping Association
PSR	passenger space ratio
PT	part-time

PTA	prepaid ticket advice
PWBLF	Prince of Wales Business Leaders Forum
RRP	recommended retail price
RTA	Regional Tourism Authorities
RTAS	Regional Tourist Associations
RTBs	Regional Tourist Boards
RV	recreation vehicle
SGIT	special group inclusive tour fare
SIC	standard industrial classification
SICTA	Standard International Classification of Tourism Activities
STB	Scottish Tourist Board
SVQs	Scottish vocational qualifications
SWOT	strengths, impotence, opportunities, and threats
TCF	travel compensation fund
TDAPS	tourism development action plans
TIC	tourist information center
TPM	ticket point mileage
TQM	total quality management
TWA	Trans World Airlines
UFTAA	Universal Federation of Travel Agents' Associations
UHT Milk	ultra-heat-treated milk
UK	United Kingdom
UKTS	United Kingdom Tourism Survey
UNESCO	United Nations Educational, Scientific, and Cultural Organization
USP	unique selling proposition
VAT	value-added tax
VDU	visual display unit
VFR	visiting friends and relatives
VR	virtual reality
WI	West Indies
WTB	Wales Tourist Board
WTM	World Travel Market
WTO	World Tourism Organization
WTO	World Trade Organization

PREFACE

Tourism refers to the movement of people across places and countries, outside their usual environment. This movement can be for a variety of purposes like entertainment, personal, professional, or business needs. In fact, it can be considered a social, economic, and cultural phenomenon, which entails the exchange of knowledge, ideas, cultures, languages, food, etc. It therefore plays a major role in boosting the economy of any country by creating jobs. Thus, aiding in the development of the infrastructures, and planting a sense of cultural exchange between citizens of the country and foreign nationals.

However, many countries in the world are solely dependent on tourism for revenue. The jobs created by tourism are spread across different sectors in the service sector but indirectly create jobs in the agricultural, communication, health, and education sectors as well. Moreover, many tourists travel to explore destination cultures, rural life, local cultures, traditions, etc., which helps local shopkeepers, small restaurants, and stores. Hence, countries that depend on tourism for the major share of their economy invest a lot in improving infrastructural developments like highways, new roads, flyovers, and tunnels which positively impact the country's development and benefit its citizens. Better infrastructure leads to the smooth and safe flow of goods and services, leading to economic prosperity.

Tourism has therefore largely impacted the transportation facilities across the globe as it is a major reason for many new airports, which have now contributed to the vital connectivity across the globe. This has helped the exposure of local businesses with access to the global market. Therefore, tourism launches the exchange of culture and tradition between tourists and citizens and, over the years, can bring a considerable change in the lifestyle, food habits, and culture of the countries. Exhibitions, conferences, and events organized for foreigners gain profits for the hosting nation.

Nevertheless, tourism can be majorly categorized into three groups; domestic tourism, inbound tourism, and outbound tourism. Domestic tourism refers to the activities when a visitor travels within his own country of residence, but outside their home of residence. Inbound tourism is when a visitor travels within the country, and Outbound tourism is when a visitor travels to a country overseas. It can be further categorized on the basis of the activity a visitor involves in their destination, which includes adventurous tourism, environmental tourism, food tourism, wellness tourism, health tourism, cultural tourism, historical tourism, recreational tourism, etc.

This book will introduce the readers to the field of tourism and the fundamentals terms of it. It is precisely designed for readers or students with no prior experience related to tourism, and bring up a diversity of fundamental topics. By the end of the book, readers will get a basic idea on tourism and the key concepts involved.

A

ABTA

Holding scheme bonding plan worked by the UK Association of British Travel Agents (ABTA), which requires all organizations to set up a bond as a state of participation. The plan covers items and administrations given by ABTA individuals. For instance, on account of ABTA visit administrator disappointment, while on vacation (excursion), the holidaymaker will be protected; on thc off chancc that the occasion has not been begun, any cash held by the ABTA visit administrator or travel planner will be returned.

ABTOT Bonding Scheme

Holding plan of the Association of Bonded Travel Organizers' Trust Ltd. worked by Travel and General Insurance Company, which covers all cash paid to travel coordinators by holidaymakers in the UK.

Abbey

A Christian strict foundation for priests or nuns, comprising of living quarters and different structures gathered cycle a congregation.

Above the Line

Business alluding to pay and use before the charge. Promoting publicizing must be paid for and the expense of which incorporates a commission paid to a promoting organization.

Aboyeur

Catering the individual in the kitchen at an eatery request from the server to the gourmet experts, and pins the server's composed request on a snare connecting with a specific table. Likewise called broadcaster.

Acapulco Document

The result of a gathering assembled in 1982 by the World Tourism Organization (WTO) in Acapulco, Mexico, to execute the points concurred by the Manila Declaration.

Academy of Culinary Arts

Catering a relationship of gourmet experts, café chiefs, and their providers, whose point is to increase expectations in food, cooking, and administration.

Accelerated Freeze-Drying

Cooking is a technique for saving food by warming it for a brief time frame, then freezing it quickly and drying it in a vacuum (Figure 1).

Figure 1. Freeze-drying.

Source: Image by Flickr.

Accessibility

The simplicity of approach of an area from different areas is one of the great elements which, along with its attractions and amenities (the 3 A's), decide how significant a region might be as a travel industry objective. Within the travel industry, availability is an element of distance from the focuses of the populace, which comprise vacationer markets, and of outer vehicle, which empowers a location to be reached. It is estimated as far as distance voyaged, the time is taken or the expense in question.

Accident

A terrible occasion that happens out of nowhere and damages somebody's wellbeing. Mishaps for the most part happen when individuals are worn out or not focusing on the thing they are doing.

Accommodation Provider

A spot that offers rooms that can be utilized by travelers or different guests, for example, a lodging, guesthouse, B, and B or inn.

Accompanied Baggage

Went with stuff is conveyed in a similar vehicle as to the traveler (and might be checked or uncontrolled things); unaccompanied things are conveyed independently as freight. In order to stay away from abundance stuff rates, it is once in a while less expensive for travelers to send a portion of their stuff as unaccompanied stuff. For instance, aircraft ordinarily convey such stuff at 50% of freight rates, which is frequently not exactly around 50% of the rate charged for abundance stuff, yet this is dependent upon specific carrier and customs guidelines.

Accompaniment

Catering a little aiding of food which is presented with a dish, for example, bread garnishes presented with fish soup or ground parmesan cheddar presented with minestrone.

Account

Finance is a record of monetary exchanges throughout some undefined time frame. For example, cash paid got, acquired, or owed. Therefore, business (in a shop) is a course of action that a client needs to purchase merchandise and pay for them sometime in the not-too-distant future, frequently at the month's end.

Activity Holiday

A conventional term for occasions (excursions) in light of a specific game, leisure activity or other interest, which empower members to seek after it as a type of dynamic diversion. It may further require not to be embraced with the end goal of working on their guidelines of execution.

Actual Flying Time

The all-out time that an airplane spends in the air during a flight, instead of its booked flight time or time spent looking out for the ground.

Added Value

A sum added to the worth of an item or administration, which is equivalent to the contrast between its expense and the sum got when it is sold. Additionally, called esteem added.

Additional Holiday (Vacation)

The term is utilized in some occasion (excursion) reviews, e.g., British National Travel Survey (BNTS), to recognize primary occasion (get-away) and others, when more than one occasion (get-away) is taken by respondents during the year, a significant pattern in most evolved nations.

Added Substance

A compound substance that is added to food to work on its appearance, smell, or taste, or to keep it from turning sour.

Add-On

An additional a discretionary thing, which is recorded in the program subtleties of a gathering or bundle visit, however, for which an extra charge must be paid.

Admission Charge

The cost to be paid prior to going into a spot, for example, to see a show or a game. Additionally, called a passage charge.

Admit

To permit somebody to go in, to say that something is right, to say that something terrible truly occurred.

Ad Valorem

'As per esteem' is a term utilized as a descriptive word related to expenses, obligations, and comparable installments, signifying that they are required with respect to the worth as, e.g., is esteem added charge (VAT), and not as a proper sum as, e.g., is regularly a flight charge.

Advance Purchase Excursion (APEX)

Exceptional return air toll presented via carriers on planned flights, dependent upon different limitations, including least development booking period and least/greatest stay necessity. It should be purchased and paid for at the hour of booking and cannot be changed or dropped without charge.

Advanced Passenger Train (APT)

A highspeed train of British Railways intended to work on conventional tracks utilizing a unique shifting system. Following quite a long while's examinations in the mid-1980s, the train was brought into administration in 1984 but removed as a result of specialized issues in 1985.

Adventure Holidays (Vacations)

Movement occasions (travels) that contain a component of an individual test, through controlled chance, trying, and additionally fervor, frequently in a difficult to reach (wild) climate. Models incorporate buckling, hang coasting, rock climbing, safaris, and wilderness boating.

Advertised Tour

A term frequently utilized freely to allude to any visit remembered for a pamphlet or different media. Nonetheless, in the USA, a promoted visit has more exact importance: it means a supported visit, which incorporates air transport and meets explicit airline requirements, to which an IT (comprehensive visit) number has been doled out, and which permits travel planners to get abrogating commission in regard of the air transport component of the visit.

Aerosol

- Little particles of fluid that stay suspended in the climate like a fog spray distributor a compartment or gadget from which fluid can be showered, in minuscule particles.
- Something can that showers out fluid as little drops (Figure 2).

Figure 2. Aerosol.

Source: Image by Flickr.

Affinity Group

A gathering who shares something for all intents and purposes, for example, an exceptional interest, or enrollment of an association or an ethnic local area.

Agency

- An office or occupation of addressing one more organization in a space;
- An office or business which organizes things for different organizations.

Agent

- An individual or organization addressing someone else or one more organization in a space;
- Someone accountable for an organization.

Agri-Food

Alluding to ventures that are associated with the mass production, handling, and investigation of food produced using rural items.

AIDS (Acquired Immune Deficiency Syndrome)

A sickness of the insusceptible framework, brought about by the HIV infection, which is sent through blood or real emissions, obliterates the body's capacity to battle diseases and is generally lethal. Moreover, full structure acquired immune deficiency syndrome illness sickness.

Air Fare Rebates (Groups)

The accompanying primary gatherings are qualified for limits off typical first, club, and economy air tolls, the size of the decreases shifting among areas and as per different models: normal vested parties, show gatherings, families, IATA traveler deal specialists on acclimation trips, motivating force visits, school parties, guests to exchange fairs. A few comparative decreases likewise apply in transport and mentor, rail, and ocean transport.

Airbridge

A covered walkway that associates an airplane with the terminal structure, so travelers can stroll onto or off the airplane without any problem. Likewise called finger 3, jetway, jet bridge.

Airbus

A brand name for a huge traveler stream airplane produced by aviation organizations from various European nations cooperating.

Airline Identification Code

A bunch of letters, which are given to all flights worked by an aircraft, for example, BA for British Airways, LH for Lufthansa, AF for Air France.

Air Mail

An approach to sending letters or bundles via air, airmail sticker a blue sticker with the words 'via airmail,' which can be adhered to an envelope or parcel to show it is being sent via air.

Air Miles

A brand name for a framework that gives individuals focuses when they buy labor and products, which they can later use to get free air travel.

Air Miss

An occurrence where two airplanes come exceptionally near one another when in the air, without causing a mishap.

Air Passes

Aircraft tickets for broad travel inside huge nations or regions, regularly accessible just to guests, who need to get them in their nations of home. Some airspaces are substantial for limitless travel in a given period; others cover a set number of flights, yet others offer limits as opposed to free travel. However, seats are dependent upon accessibility; a few aircraft offer booking affirmed seats; others reserve as it were. Air passes are presented by most huge US and Canadian carriers, some covering the two nations; a few covers Australia, the New Zealand, and the Pacific islands, and blends of these nations.

Airport Code

A bunch of letters that are utilized to distinguish a specific air terminal, for example, LHR for London Heathrow or CDG for Charles de Gaulle.

Air Travel Organizer

An individual who makes 'accessible, as a head or a specialist, convenience for the carriage of people or freight in any area of the planet' requires a permit to work by the UK Civil Aviation Authority (CAA) (Civil Aviation Act, 1971). This is a more extensive term than visit administrator and incorporates, for instance, air specialists going about as specialists for carriers. Moreover, the permit is the air travel organizer's license (ATOL). Hence, the permit shows that the coordinator of sanction flights has stopped security with CAA to protect the holidaymakers' cash in the event of the organization's disappointment.

Airport Transfer

A vehicle administration to take travelers from an air terminal to an inn, gathering focus or another point, or from that point back to the air terminal.

Air Travel Organizers' License

A permit that must be held by any organization or individual contribution bundles occasions or contract flights and incorporates a cling to safeguard voyagers assuming the organization goes into liquidation.

Air Passenger Duty (UK)

Charge payable on tickets for flights withdrawing from UK air terminals presented in 1995 and corrected from April 2001 and again November 2002. It differs between trips inside and outside Europe and furthermore among economy and higher-class flights. The sum is remembered for the cost of a ticket or occasion (get-away) by the administrator and represented to the Treasury.

Airline Clubs

Unique clubs are made by the most significant aircraft for continuous explorers, offering different honors and administrations, for example, the utilization of extraordinary air terminal parlors. Moreover, enrollment of clubs run by US carriers is typically open to anybody arranged to pay a decent yearly participation charge. Hence, participation of clubs run by different carriers is frequently free yet confirmation is limited to that thought about important enough clients.

Airport Capacity

Air terminal limit is comprised of the limits of:

- The terminal (the number of travelers each hour traveling through such pieces of the terminal structure as security, migration, and customs);
- The cover (the number of airplanes dealt with each hour, which relies upon the quantity of stopping stands and the airplane adjusting capacity);
- The airplane developments (the number of developments each hour that can be upheld by the air terminal).

Alcohol

An unadulterated vapid fluid that is framed by the activity of yeast on sugar arrangements and structures a piece of beverages like wine and whisky.

Alcopop

A beverage made and sold monetarily, that is a combination of soda pop, for example, lemonade, and liquor.

Algae

Small plants living in water or in sodden circumstances, which contain chlorophyll and have no stems or roots or leaves (Figure 3).

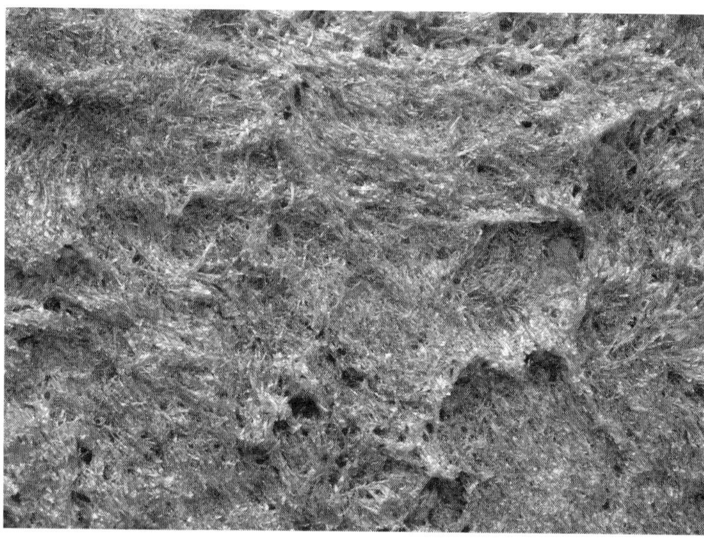

Figure 3. Algae seaweed.

Source: Image by Pixabay.

Alimentary Canal

A cylinder in the body going from the mouth to the butt and including the throat, stomach, and digestion tracts, through which food passes and is processed.

Alternative Tourism

The term is commonly used to allude to types of the travel industry that look to stay away from unfriendly and upgrade positive social, and natural effects and saw as a choice to mass the travel industry. Typically portrayed by limited scope; individual, autonomous, or little gathering movement;

slow, controlled, and directed advancement; as well as an accentuation on movement as the experience of host societies and on the upkeep of customary qualities and social orders. Likewise alluded to differently as fitting, green, capable, or delicate the travel industry.

All-In Price

A cost which covers every one of the things in a buy, for example, the merchandise, in addition to conveyance, charge, protection, and so forth, or all things in a visit like travel, inn convenience, and dinners.

Allowed Time

The paid time that the administration concurs a worker can spend on rest, cleaning, or suppers, yet at the same not working.

Allocentric/Psychocentric

Terms to depict sorts of vacationers, ascribed to the travel industry scientist Stanley Plog. An allocentric is an explorer with an inclination for investigation and curiosity, consistently looking for new objections. Moreover, a psychometric is unadventurous, needs security, looks for the recognizable and has an 'attempted and tried' objective. In the middle of these limits is the mid centric who has some of the two qualities and makes up the heft of the market.

All-Terrain Boarding

A type of skating where the rider goes over a wide range of territory, particularly down mountain slants.

Altitude Sickness

A condition brought about by diminished oxygen in the air above heights of 7,000 to 8,000 feet (3,600 meters). Side effects incorporate cerebral pains, shortness of breath, weakness, queasiness, and expansion of the face, hands, and feet. Additionally, called mountain affliction.

Amenity Center

A structure lodging different diversion office, like a film, sports lobby, exercise room, amphitheater, and pool.

American Express

An organization offering a movement administration, secured checks, charge cards, and numerous different administrations around the world.

American Service

A way of laying a table, where every visitor is given cutlery, a side plate, a napkin, glasses, and now and then an espresso mug and saucer.

Amoebic Dysentery

A mostly tropical type of loose bowels that is brought about by microorganisms that enter the body through sullied water or unwashed food.

Animator

Someone utilized to sort out stimulations or different exercises for visitors in a lodging or occasion resort or for travelers on a boat.

Apartment Hotel

An inn is framed by a progression of outfitted rooms or suites and where all ordinary inn administrations are given, albeit every suite will have its own kitchenette.

APEX Fare

A uniquely modest air toll that you should book a specific period of time before the flight and that you can change or drop by paying an additional an extra charge.

Apron Congestion

A circumstance where an excessive number of planes attempt to utilize the cover at an air terminal, bringing about more slow turnaround times.

Appropriate Tourism

The sort and size of the travel industry which is thought of as reasonable for an area taking into account its financial, social, natural, and different circumstances. The term is additionally utilized as an equivalent for electives in the travel industry.

Arctic

The region inside the Arctic Circle (66°32' North), inverse the Antarctic; additionally, as a descriptive word, of that area; e.g., arctic environment. The district is lacking in a travel industry objective and is at present visited fundamentally by researchers, understudies, and adventurers. Be that as it may, particularly northern Canada and northern Scandinavia are appealing for those looking for the wild and may get more guests in store for experience occasions (get-aways) and summer travels (Figure 4).

Figure 4. The arctic ice pack with pressure ridges at the geographic north pole.

Source: Image by Wikimedia Commons.

Area of Outstanding Natural Beauty

A district in England and Wales that is anything but a National Park yet is viewed as adequately alluring to be saved from overdevelopment.

Area Tourism Companies (ATC)

The region structure of three bodies covering South and West, Mid, and North Wales, with nearby power and the travel industry support and working under an agreement with the Wales Tourist Board (WTB) to create, advance, and direction the travel industry in their particular region of the Principality.

Assisted Passage

An excursion for a worker to an unfamiliar country is part of the way paid for by the public authority of that country to energize the movement.

Atrium

Encased covered person on foot space frequently framing the entryway or an inside patio of huge structures, including hot.

Attendance Factor

An unpleasant proportion of the adequacy of a recreation office or administration is determined by separating absolute attendances by the populace inside its apparent catchment region. This might be contrasted and a public standard got from such sources as the general household survey (GHS) in Great Britain (GB).

Attribution Theory

A way to deal with understanding how individuals make sense of negative encounters or hazardous occasions. In the travel industry, the hypothesis might help, e.g., with sightseers' protests.

Authenticity

The nature of being certifiable, genuine, or valid, instead of recreated, imagined or counterfeit. Applied in the travel industry specifically to legacy destinations and to occasion attractions, while alluding, e.g., to structures and objects of craftsmanship or to services and exhibitions.

Automated Immigration Lane

Computer-controlled movement systems worked first as a worldwide investigation in various areas, for example, air terminals and furthermore the US-Mexico line, which utilizes biometrics and recognizes individuals from the qualities of their hands. Therefore, assuming the trial is effective, it is conceived that the innovation could, at last, supplant visas.

Automated Teller Machines (ATMs)

Otherwise called cash focuses or cash allocators. They are offices presented by banks and building social orders to current record holders to acquire cash up to a concurred sum and such different administrations as the equilibrium of one's record. This is done through the embedding a help card in the machine and scratching a PIN number.

Automobile Club

Individual enrollment associations give side of the road and different administrations to drivers, like the Automobile Association (AAA) in Great Britain (GB), or the American Automobile Association (AAA).

Avalanche

The fast plummet down steep slants of masses of snow, ice, or rock, generally abruptly. The term is usually applied to snow when it very well might be set off by weighty falls in winter or by unexpected spells of warm weather conditions in spring.

Most torrential slides follow ordinary torrential slide tracks and to that degree are unsurprising concerning area. However, torrential slides might happen now and again out of the blue in different regions. Torrential slides are a specific danger to climbers and skiers, particularly those skiing off-piste.

Axes

A PC reservation framework (CRS) possessed by Japan Air Lines and Korean Airlines with a collaboration concurrence with Saber.

B

Baby-Listening Service

Help is given by a lodging, with a little amplifier to put over a child's bed, so the guardians can hear in the event that the child cries when they are not in the room.

Backpacking

Casual amusement, especially in hilly and wild regions, by those conveying their stuff and food supplies, by and large in backpacks, and depending on camping cots, tents, or hovels for convenience. Consequently, a hiker, a voyager occupied with exploring who will, in general, be normally youthful and on a low spending plan (Figure 5).

Figure 5. Backpacking and trekking life at Mustang, Nepal.

Source: Image by Wikimedia Commons.

Baby-Sitting Service

Help is given by an inn, where a sitter comes to the lodging to take care of a child when the guardians are out of the room.

Backhander

An amount of cash given covertly and illicitly to someone to convince the person in question to help you (casual).

Back-of-House Services

Administrations that are situated in the back piece of lodging. For example, cleaning, and giving supplies to the café and bar. Likewise got back to of-the house administrations.

Back-to-Back

A succession of gathering appearances and takeoffs were organized in order to expand the use of the method for transport or convenience. For instance, a contract airplane ships a gathering of travelers to an objective on its outward flight and gets back with another gathering, or lodgings cleared by a gathering of leaving visitors are involved that very day by another gathering showing up.

Back-to-Back Arrangement

A course of action covering travel and lodging convenience, coordinated with the goal that when one gathering shows up they consume similar rooms as the past gathering, who leave by a similar plane.

Back-to-Back Ticketing

The training (against IATA guidelines) of purchasing numerous return passes to keep away from guidelines requiring Saturday night stay at the objective in order to fit the bill for a diminished toll. Subsequently, e.g., to visit Paris from London two times in two weeks (flying out one work day and returning the following each time) could be covered by purchasing two return tickets (both over Saturday evenings), one London-Paris and the other Paris-London, and 'blending' the tickets.

Bacteria

Submicroscopic organic entities help in the deterioration of natural matter, some of which are forever present in the digestion tracts of creatures and can separate food tissue, and some of which cause illness.

Bag

A delicate holder made of plastic, fabric, or paper, and utilized for conveying things, a delicate case for conveying garments while voyaging.

Baggage Allowance

A measure of things, not entirely settled by weight, a number of pieces and additional aspects, shipped by a transporter 'free.' For instance, under the weight framework utilized by numerous aircraft, this is ordinarily 30 kg (66 lb) for five stars and 20 kg (44 lb) for every economy traveler.

Baggage Check

- An assessment of travelers' things to ensure it contains nothing risky or unlawful;
- A receipt given to a traveler for stuff which has been checked in, and which is normally stapled to the travelers' ticket counterfoil.

Baggage Check-In

Where air travelers have their packs and bags gauged and hand them over to be placed on the airplane.

Baked Alaska

A pastry made of frozen yogurt covered with meringue, prepared in a broiler for a brief time frame to cook the meringue, eaten before the frozen yogurt dissolves.

Baked Beans

Haricot beans, cooked in a pureed tomato, customarily heated on the stove with bits of pork and molasses, and called 'Boston prepared beans,' yet presently more for the most part accessible in jars (Figure 6).

Figure 6. Boston baked beans.

Source: Image by Pixabay.

Baked Potato

A potato cooked 'in its coat,' that is, heated in a stove without being stripped, then, at that point, served cut open, with margarine or different fillings like cheddar, hacked ham, prepared beans, stew, and so forth. Likewise called jacket potato.

Balance Brought Forward

An equilibrium is placed in a record toward the finish of a period and is then taken to be the beginning stage of the following time frame.

Balance of Payments

A record of a country's exchanges with the remainder of the world, i.e., with different nations and worldwide establishments. The fundamental divisions are the ongoing record and the capital record. The ongoing record is sub-isolated into noticeable things (equilibrium of exchange, i.e., commodities, and imports of products) and invisibles (administrations, profit from ventures, private exchanges, and government exchanges). Moreover, the capital record incorporates cash streams for speculation, awards, and advances. Global travel spending in the nations visited is displayed as a different imperceptible thing; different exchanges because of movement and the travel industry are lowered with different things.

Baltic States

The term is given to the nations on the eastern side of the Baltic Sea – Lithuania, Latvia, Estonia – which were free between the two World Wars, attached by the Soviet Union in 1940, and have been autonomous again beginning around 1991. Therefore, the Baltic States have numerous attractions, yet numerous offices are still to be created for the global travel industry.

Bar Code

An arrangement of lines imprinted on an item can be perused by a PC to give a reference number or cost. Additionally, called Universal Product Code.

Bard

Put a segment of fat or greasy meat, like a rasher of bacon, over the meat to keep it from drying out while cooking.

Bareboat Charter

A means of sanctioning a boat where the proprietor gives just the boat, however not the team, fuel, or protection.

Barometer

An instrument for estimating environmental tension, and in this way for determining the climate.

Batter

A meager fluid combination of flour, eggs, and milk, is utilized, for example, for making hotcakes or frog in-the-opening, or for covering fish prior to broiling.

Beaujolais Nouveau

Beaujolais wine which has recently been made is sold from November onwards of the year wherein the grapes are picked.

Beach

An area of sand or shingle on the shore of a lake or of the ocean; W when the last option, fundamentally among low and high watermark, however frequently stretching out above highwater and upheld by precipices, rises, and vegetated land. On a rough shore, in bayous between headlands, seashores are naturally arcuate in shape and under a mile long, however, on delicate coasts they can be significantly longer and straighter, e.g., Ninety Mile Beach in New Zealand. Seashores are significant. Vacationer/guest fascination, both by their own doing and for water-based exercises.

While seashores are, inside the flowing reach, generally self-purging, they, and the seaward waters in thickly settled regions, have been incredibly impacted by contamination, litter kept by clients, decline unloaded seaward, metropolitan sewage, and oil slicks. Moreover, the European Union (EU) is endeavoring to advance clean seashores by an arrangement of assessment and the honor of Blue Flags for those that accomplish the necessary norm.

Bed, Breakfast, and Evening Meal

A levy in an inn or guesthouse, covering a night's convenience, breakfast, and dinner taken in the evening.

Bedside Panel

An electronic gadget put on a visitor alongside the table, with controls for warming, lighting, cooling, programed drapery pulling, informing, and so forth.

Bed Tax

Charge demanded by focal or nearby government or one more organization on remaining guests gathered at the spot of stay, for of raising income; at times the returns are applied to the travel industry purposes. Might be likewise called lodging, or room charge.

Beefburger

A round, level cake of minced hamburger, barbecued or seared and ordinarily served in a toasted bread roll. Likewise called a burger, a cheeseburger.

Beermat

A little cardboard mat, typically with a notice for a brewery on it, put under glass to safeguard a surface in a bar or eatery.

Beer Pull

One of a progression of tall handles organized on the bar of a bar, which draw brew up from barrels in the basement when pulled towards the barkeep.

Below-the-Line Advertising

Publicizing that isn't paid for; for example, work by staff monitoring a presentation, and for which no commission is paid to the promoting organization.

Berry

A little beefy seed-proving to be fruitful of a shrubbery, generally with many seeds in a similar natural product, and the seeds encased in a mash.

Best-Before Date

A date stepped on the name of a food item, is the keep going date on which the item is destined to be of good quality.

Best Practice Forum

A framework supported by the UK government to energize exclusive expectations of incredible skill in the lodging, catering, and the travel industry enterprises.

Beverage Cycle

The cycle by which drinks pass from provider to lodging or café and afterwards to a definitive shopper.

Bicycle Path

An extraordinary way for bikes to ride on, either by the side of a street or as a component of the asphalt.

Billiards

A game includes hitting balls with a long pole on a smooth table covered with green material, the item being to hit a white ball so it sends a chunk of one more tone into one of the 'pockets' at the edge of the table (Figure 7).

Figure 7. Billiards.

Source: Image by Pixabay.

Biotechnology

The utilization of innovation to control and consolidate different hereditary materials to deliver living life forms with specific qualities.

Blackcurrant

A little round developed dark berry, eaten cooked in predicaments and pies, utilized additionally in making soda pops and mixers.

Black Tie

Alluding to a proper evening feast or gathering at what men wear a necktie and supper coat, both generally black.

Black Market

Unlawful exchanges in scant items, authoritatively controlled products, or monetary standards. Whenever monetary standards are to a critical degree traded in the underground market throughout worldwide travel/the travel industry, it is hard to show up at dependable assessments of global the travel industry receipts and consumptions; this is especially the situation when the appraisals depend on such backhanded strategies as bank detailing instead of guest overviews.

Block Spacing

An allotment by one aircraft to one more of various seats on a portion of its flights, which the carrier offers to the voyaging public through its own showcasing and appropriation framework. A square dividing arrangement is utilized while the designating carrier has a spare limit and the aircraft to whom the seats are allotted is for reasons unknown unfit to serve an air terminal.

Blind-Tasting

A party where a progression of things, like cheeses or wines, are tasted without the testers having the option to see their names.

Blue Channel

The exit from customs through which you pass assuming you are showing up from another EU country and are not bringing in merchandise which is obligated to the obligation.

Boarding Card

A card is given to travelers who have checked in for a trip to permit them to load onto the plane or a card is given to travelers getting onto a boat.

Booking Charge

Cash is paid to an office for their administration notwithstanding the expense of the ticket when you purchase a ticket through them. Additionally, called booking expense.

Booze Cruise

An excursion across the English Channel to purchase cocktails in a nation where they cost short of what they do in the United Kingdom (UK; casual),

Border

A line that denotes where one nation or district closes and another starts.

Border Crossing

A put on the boundary between two nations where individuals can cross and where there are visa controls and customs posts.

Bouncer

Someone at the entryway of a club, whose work it is to keep nonmembers or other undesirable visitors from entering.

British National Travel Survey (BNTS)

A yearly example enquiry did by private meetings discontinuously during the 1950s and afterwards, every year beginning around 1960 by a statistical surveying organization for the British Tourist Authority (BTA) to layout the degree of occasions (get-aways) away from home by British occupants and data about special times of the year they take in Britain and abroad.

Nevertheless, for a period (1985–1988) likewise alluded to as British tourism survey yearly (BTS-Y). Synopsis results show up in the digest of tourist statistics distributed by the BTA. To be recognized by National Travel Survey.

British Tourism Survey (BTS)

Name given to two separate example reviews of British inhabitants' travel industry somewhere in the range of 1985 and 1988: (I) A yearly occasion (get-away) study in late pre-winter, known as British tourism survey yearly (BTS-Y), yet until 1984 and again starting around 1989 as British National Travel Survey (BNTS); and (II) a month-to-month overview of the travel industry as far as purposes, known as British tourism survey monthly (BTS-M), however previously called British home tourism survey (BHTS) and in 1989 supplanted by United Kingdom Tourism Survey (UKTS).

British National Rail Apex

A return ticket following aircraft idea and wording and accessible on InterCity trains from specific stations in Great Britain (GB) with the fixed allotment of seats, offering a huge saving money on ordinary charge in addition to the free situate reservation. Moreover, there should be reserved somewhere around seven days ahead of time and return date affirmed simultaneously; travel should be on the train named on the ticket; there is a crossing out expense.

Bumping

A circumstance where somebody replaces one less notable individual in an eatery or on a plane.

Bungee-Jumping

A critical experience of the travel industry movement whose cutting-edge form began in New Zealand, however, professed to have its beginnings in the land jumping actually arranged on the island of Pentecost in Vanuatu. There it is performed by men making a plunge from a 70 ft tower with plants appended to their lower legs to dampen the fall (Figure 8).

Figure 8. Rocket bungee jumping.

Source: Image by Wikimedia Commons.

Business Lounge

An exceptional parlor for business class travelers, with agreeable seats, bistro, and so forth, yet additionally some business offices, for example, Internet joins fax machines and secretarial assistance.

Bunk

Two-level bed, normal in youth lodgings, and boats; henceforth bunkhouse, an office giving such convenience, now, and then as a piece of a bigger office, for example, an inn or café offering extra administrations, including suppers and rewards, to bunkhouse clients, found, e.g., in Scotland.

Busboy/Busgirl/Busser

American expression for collaborator server or server, who pours water, gathers up plates and cutlery but doesn't take requests or serve the food.

Business Attire

Semi-formal dress of matching suit or coat with a shirt and tie for men and day dress or suit for ladies.

Business Class

A class of transport, typically carrier, administration among first and economy class, i.e., more affordable than the previous and more agreeable than the last option, with different exceptional conveniences. Otherwise called Club Class on certain carriers.

Business Cycle

Kind of vacillation in monetary action around the more drawn-out term pattern seen in industrialized nations, comprising of general development followed by an also broad downturn. Reflected in the degree of public pay, work, and different totals, the change is intermittent. However, the range from one top to another or box to box of the cycle fluctuates, commonly something like five years or more. Moreover, the general cycle impacts the interest for movement, the travel industry, and friendliness administrations, as well as the other way around, albeit various structures vary in the degree and furthermore in the delay of the impact.

Business Travel/Tourism

Outings and visits made by representatives and others over their work, including going to gatherings, meetings, and presentations.

Buyers' Market

A business opportunity for labor and products where purchasers are in a solid bargaining posture, generally as an outcome of an abundance of supply over interest, portrayed by falling costs. In movement, the travel industry and friendliness, a purchasers' market exists when there is, e.g., overcapacity of aircraft seats or lodgings.

Bypass

- Road cycle a town giving an elective course to through traffic. Called a freeway in the USA.
- In the USA, selling by transport and visit administrators direct to a client without the utilization of travel planners, is known as immediate selling in the UK.

C

Cab

A two-wheeled one-horse carriage with an enormous hood, yet these days utilized all the more broadly for a vehicle for recruit by the overall population with a driver and a meter for enrolling the admission, as an equivalent for a taxi.

Cabin

- A wooden hovel, utilized by trackers or skiers, likewise the resting region in a ski chalet;
- A different space for a traveler on a boat.

Cabin Class

A class of convenience on some traveler transports that is lower than top-notch and higher than vacationer class.

Cable Car

- A vehicle which goes up a mountain, holding tight a wire link;
- US (in San Francisco) is a sort of cable car which is pulled by a metal link set in a station in the street.

Caesar Salad

A plate of mixed greens made with lettuce, bread garnishes, Parmesan cheddar, and anchovies, with an egg-based dressing.

Café Set

A piece of café gear which warms the water and makes steam, for getting ready hot beverages.

Cafeteria

Self-administration cafés are frequently situated in a production line, office block or somewhere else of work for the utilization of staff or in an instructive foundation for the utilization of understudies and staff (Figure 9).

Figure 9. A cafeteria.

Source: Image by Flickr.

Cafeteria Service

A way of serving food is where the client takes a plate and helps oneself to be hot or cold food from a smorgasbord and pays for it at a till as the individual in question leaves the smorgasbord.

Cake Fork

A little fork with two the prongs consolidated, utilized for eating cake or baked goods. Additionally, called a cake fork.

Calamine Lotion

A salve, in view of zinc oxide, assuages skin aggravation, for example, that brought about by burn from the sun.

Calendar

A book or set of pieces of paper showing the days and months in a year, frequently delineated with a progression of pictures.

Camembert

A little round delicate French cheddar that has a palatable white skin and becomes more grounded in flavor and milder in the middle as it matures.

Camper Van

An engine parade outfitted with bunks, kitchen hardware and furniture, in which individuals can cruise all over and park to remain for the time being.

Camping Site

A region exceptionally set up for setting up camp and trains, with stamped places for tents and public latrines, washrooms, and different offices.

Canteen

- A private self-administration café in an office square, production line or comparative structure;
- A container containing blades, forks, and spoons;
- A convenient cup for water.

Canyoneering

The game of walking through gulches, utilizing abilities, for example, abseiling, swimming, climbing, and boating.

Caper

The flower bud of a Mediterranean bramble is cured and utilized in sauces or as a trimming for fish and meat.

Captain

Someone accountable for a boat or airplane.

Caravan

A gathering of dealers, pioneers or others voyaging together through deserts and other 'void' regions in the East and in northern Africa. See additionally caravanserai. (b) Accommodation vehicle for extremely durable, occasional, or transitory occupation, which meets necessities for development and utilization of street vehicles. Moreover, essential qualifications are between trailer (towed) and engine (self-propelled) convoys, additionally called campers and in the USA RVs. Hence, band destinations or parks are isolated by their utilization into long-lasting, static, and visiting locales, giving extremely durable puts of home and second homes from one viewpoint and brief momentary offices for occasion (excursion) use then again.

Caravanserai

An enormous hotel with a focal patio, found in a few eastern nations and utilized via trains crossing the desert.

Car Insurance

The matter of guaranteeing a vehicle, the driver, and travelers in the event of a mishap. Additionally, called engine protection.

Carnet

A tradition pass is given by motoring associations to their individuals, approving brief importation of engine vehicles, trailers, and trains into nations without installment of obligations. A carnet is additionally utilized for impermanent obligation-free importation of specific different products, e.g., limited time material of no business esteem, masterpieces, and capital hardware for transitory use in shows.

Carob

A palatable powder with a taste like that of chocolate, produced using the seeds and units of an evergreen tree.

Carousel

A gadget at an air terminal comprising of a turning stage or a huge round belt, where the stuff of showing up travelers is put by stuff controllers so the travelers can find it and remove it.

Carpaccio

A dish of crude hamburger cut daintily, saturated with olive oil and lemon squeeze and prepared (Figure 10).

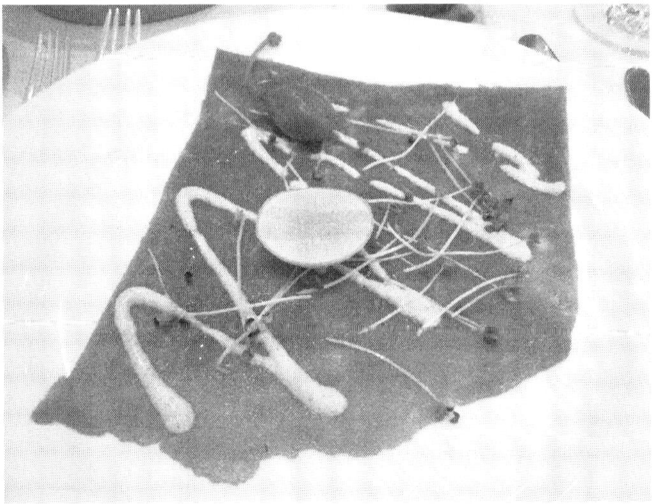

Figure 10. Carpaccio.

Source: Image by Flickr.

Carriage

A mentor for travelers on a train, the demonstration of shipping merchandise starting with one spot then onto the next, or the expense of doing this.

Carrier

Any individual or association that embraces the movement of products or potentially individuals for recruitment. The lawful term normal transporter suggests a commitment to ship subject to specific circumstances (e.g., on account of rail routes), as unmistakable from private transporter who makes no broad proposal to general society. Moreover, the term assigned transporter alludes to a carrier recognized in a two-sided consent to work administrations between two nations.

Carte Du Jour

Menu of the day, a rundown of extraordinary dishes ready for the afternoon and not recorded in the printed menu.

Cartouche

A round or oval piece of paper that is put on top of food while it is being cooked in fluid, for example, in a goulash, to keep the strong fixings lowered.

Carver

- Someone who cuts meat, frequently an extraordinary server in a café, who cuts a joint brought to the side of a table on a streetcar;
- A seat with arms, set at the top of a supper table.

Carvery

An eatery where hot dish dinners are presented at the smorgasbord table for the client to cut his/her own piece or to be served by a gourmet expert and afterwards to help himself/herself to vegetables and backups.

Carving Station

US a huge, frequently warmed, platter on which a major joint of meat can be cut for serving to visitors in an inn or café.

Carving Trolley

An extraordinary streetcar with a joint of meat in a warming dish, and a level surface for cutting it, which can be wheeled from one table to another in an eatery.

Case

A crate with a handle for conveying garments and individual possessions while voyaging, a cardboard or wooden box for pressing and conveying merchandise.

Cabana

A room or construction near the ocean or by the pool isolated from the primary lodging building may yet require not to be outfitted with beds.

Cabin Cruiser

A power-driven vessel with at least one lodge giving living and resting convenience and, in some cases, additionally cooking offices, and utilized for relaxation purposes.

Cable Car

A method for transport up the mountain utilized primarily for touring and skiing, moved by an unending link.

Cabotage

In delivery, beachfront exchange, or transport between ports of a similar country. In common aeronautics, travel, and transport on homegrown courses or between regions of a sovereign state, including its provinces and conditions. Subsequently, e.g., are not just London-Glasgow and New York-Los Angeles, yet in addition, London-Gibraltar, and Los Angeles-Honolulu are cabotage courses. Peaceful accords don't reach out to cabotage courses.

Thus, occupants of a nation might be qualified for unique diminished tolls, known as cabotage admissions. Cabotage privileges allude to one side of the country's transporter to convey travelers between two focuses in another country.

Café Establishment

Giving food and rewards to utilization anywhere nearby to the overall population. Generally, a little unit with a restricted menu and no alcoholic alcohol accessible, in some cases open just during the day yet not in the evening.

Café Complete

An early in the day or evening nibble with espresso, once in a while additionally used to portray Continental breakfast.

Campaign

As utilized in showcasing, the term portrays any coordinated program of activity in the advancement of labor and products to accomplish explicit goals. Each mission is typically arranged throughout a predetermined timeframe and has a typical topic or message.

Campus Holidays

Occasions in light of understudy convenience on college grounds during get-aways, frequently movement occasions for sure might be viewed as unique interest occasions (get-aways).

Canal

A fake conduit is built to interface streams, lakes, and different streams for: (I) inland route; or (II) water supply and water system. Significant instances of the previous incorporate the Panama Canal and the Suez Canal; various waterways in the southwest USA and in Pakistan represent the last option. Long tight boats with living convenience utilized on waterways are known as trench boats (Figure 11).

Figure 11. Bhakra main channel, Punjab, India.

Source: Image by Wikimedia Commons.

Canine Corps

Canines having a place with US customs and the Department of Agriculture used to track down drugs and other booty things at worldwide air terminals.

Canyon

A Spanish expression portraying a profound, steep-sided valley, with a stream streaming at the base, normal for the level country under bone-dry and semi-bone-dry environments.

Capacity Limitation Agreement

An arrangement between transporters, ordinarily aircraft, specifying the most extreme ability to be presented by every transporter on a specific course.

Capacity Tone-Miles

A proportion of transport yield is determined as the result of the airplane payload (estimated in short huge loads of 2000 lb.) and the miles flown.

Capacity Ton – Kilometers

A proportion of transport yield is determined as the result of the airplane payload (estimated in metric tons) and the kilometers flown. Otherwise called accessible ton-kilometers (ATK).

Captain's Table

Lounge area table on a boat, which travelers are welcome to impart to the commander during the journey.

Car Hop

An American expression for a worker:

- Who leaves vehicles of showing up visitors and returns them on takeoff; the training is known as valet leaving;
- Who serves clients in their vehicles in a drive-in eatery

Car Occupancy

The number of individuals per vehicle visiting a vacationer/guest fascination or office; a significant measurement with suggestions for the arrangement of the vehicle leaving.

Caravan

A gathering of vendors, pioneers or others voyaging together through deserts and other 'void' regions in the East and in northern Africa. Convenience vehicle for extremely durable, occasional, or brief occupation, which meets prerequisites for development and utilization of street vehicles. Therefore, essential differentiations are between trailer (towed) and engine (self-propelled) troops, likewise called campers and in the USA RVs. Parade locales or parks are separated by their utilization into extremely durable, static, and visiting destinations, giving long-lasting puts of home and second homes from one viewpoint and transitory transient offices for the occasion (excursion) use then again.

Caribbean Tourism Statistics

Measurements of individual nations as well with respect to the district, in general, are distributed in a yearly report by the Caribbean Tourism Organization.

Carousel

- Revolving component from which travelers guarantee really look at things at air terminals and other vehicle terminals.
- Food and refreshment self-administration, in which food, as well as the drink, is shown on a few rotating round racks at various levels; from that point the client puts his/her picked things on a plate.

Carrying Capacity

In the travel industry, the greatest limit of a site or region is to support vacationer movement without decay in the nature of the guest experience of the climate. Consequently, conveying limit might be believed to have physical, social (perceptual) and ecological aspects and is ordinarily communicated as far as a given number of simultaneous clients of, e.g., a memorable fascination, oceanside or resort.

Catamaran

Initially portraying a heap of two boats secured next to each other, these days the term is typically utilized for a twin frame boat or boat, intended to slice through the waves. For example, SeaCat worked as a ship by Hover speed between the English coast and the Continent of Europe.

Catchment Area

With regards to travel, the travel industry and neighborliness, a region from which most the clients of a fascination, office or administration are drawn. Many, like theaters or pools, are normally in the neighborhood, with most clients coming from inside a couple of miles. Moreover, many amusements stops and significant recreation communities have local catchment regions; a few significant attractions, like Alton Towers in England, have public catchment regions.

Caving

A general term which has procured unique significance as dynamic entertainment investigating caves, especially in limestone regions, as in the Peak District National Park in England, where there are interlinked frameworks of caves. Issues can happen from the abrupt ascent in the degree of water in the sinkholes, which might bring about cavers being caught or suffocated.

Caverns might be likewise vacationer/guest attractions by their own doing, similarly to the Carlsbad Caverns in the USA, where the fundamental cavern has a most extreme level of 285 feet.

Caveat Emptor

A legitimate articulation signifying 'let the purchaser be careful,' i.e., an individual is accepted to practice the presence of mind while purchasing and not being qualified for the assurance of the law, if he/she neglects to do as such. Therefore, in most evolved nations the standard is currently of restricted importance because of customer insurance regulation, which is especially pertinent in movement. Thus, the travel industry and accommodation, where the purchaser is many times not in that frame of mind to review the item before buying.

Chamber of Commerce

Intentional charitable making association of organizations in a town or region to secure and advance their inclinations through portrayal and offering types of assistance to individuals as well as advancing neighborhood financial turn of events. Likewise, in certain nations, they fill the roles of a nearby vacationer board, particularly the travel industry advancement. Hence, offices of Commerce are to be recognized from offices of exchange, regularly nearby relationships of retailers with additional restricted purposes, and furthermore from exchange affiliations.

Charter

The recruit by agreement of the entire or part limit (part sanction) of an airplane, boat, train, or transport; when a few administrators share a contract, this is known as parted sanction. Single or a few oddball plans are known as impromptu sanctions, normal excursions contracted for as series sanctions; on account of time contract an administrator has a restrictive utilization of the vehicle all through the time of the contract. In movement and the travel industry contracting has accepted specific importance regarding comprehensive visits via air, which were liable for a large part of the development of global travel/the travel industry following the Second World War.

Check-In Time

Term most normally used to mean the most recent time by which travelers are expected to report at the air terminal before flight takeoff. The span between registration and takeoff times typically varies among homegrown and worldwide flights, yet in addition among planned and sanction administrations to a similar objective. Moreover, the term is additionally utilized corresponding to different types of transport and to some degree distinctively in lodgings where the most recent registration time indicates how late a saved room will be held before it is let to another visitor, except if the inn is informed of late appearance.

Circle Pacific Fare

Exceptional air passage presented by a few Pacific transporters, which permits travelers to travel to Pacific Rim objections including Australia, Asia, North, and South America, generally incorporating four visits with extra visits accessible at an additional a charge.

Circle Trip

A trip including more than one destination and return to origin, as distinct from a return or round trip, a trip from one point to another and return.

City/Airport Codes

Three-letter location identifiers assigned and published by the International Civil Aviation Organization (ICAO) for use in timetables, ticketing, and other communications (Figure 12).

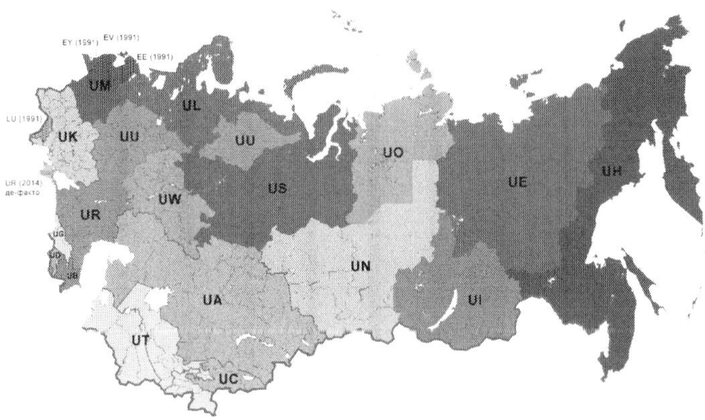

Figure 12. The diagram shows the original geographical attachment of ICAO aerodrome codes with U prefix (originally assigned for the USSR). The dashed regions later changed codes.

Source: Image by Wikimedia Commons.

City Terminal

Airline office located in town or city center away from the airport from which passengers can normally obtain transport to the airport, and sometimes also complete some or all check-in procedures for their flight. Also called air terminal.

Collective Tourism Establishment

An accommodation establishment providing overnight lodging for the traveler in a room or some other unit, but the number of places it provides must be greater than a specified minimum for groups of persons exceeding a single-family unit and all the places in the establishment must come under a common commercial-type management, even if it is nonprofit making. Hence, collective establishments include hotels and similar establishments, specialized establishments, and other collective establishments [World Tourism Organization (WTO)].

Code of Conduct

A set of guidelines laying down standards to which members of a profession or association are expected to adhere in the exercise of their activities. For example, in the UK, the Association of British Travel Agents (ABTA) provides codes of conduct for retail travel agents and for tour operators, and individual members of the Hotel and Catering International Management Association (HCIMA) are required to observe the Association's code of conduct. Moreover, the agreement between the International Hotel and Restaurant Association (IHRA) and the Universal Federation of Travel Agents' Associations (UFTAA) on hotel contracts represents a bilateral Code of Practice at international level. In recent years a number of Codes of Practice have been developed for the regulation and control of the environment, e.g., by the Pacific Asia Travel Association (PATA).

Common Interest Travel/Tourism

A major segment of travel and tourism to be distinguished from holiday (vacation) and business travel/tourism, in which the visitor and the visited significantly share the common purpose of the trip/visit, e.g., visiting friends and relatives (VFR), education, religion. It is therefore not relatively price-sensitive and not readily influenced by promotion, common interest travel/tourism often includes a relatively long stay but limited use of commercial facilities and services in the place visited.

Connecting Flight/Train

The subsequent flight or train service that will continue the passenger's journey to the next destination. In air transport, a maximum time between connecting flights is sometimes laid down in order to prevent the abuse of rules relating to stopovers. A passenger may need to check-in again, before joining a connecting flight.

Connecting Time

Although applicable in most forms of transport, of particular importance in air travel, where it is the minimum specified time to be allowed for a passenger between arrival on one flight and departure on a connecting flight. The time varies between airports, domestic, and international flights, online and offline, and according to other circumstances.

Connectivity Index

Measure of internal accessibility of a region based on established routes composed of links connecting nodes, which form a transportation network; the higher the level of connectivity, the better for tourism.

Corridor Train

- A train consisting of coaches with corridors and individual compartments, as distinct from open plan coaches.
- A train service in parts of Continental Europe, which leaves the country of origin and runs through another country without stopping before returning to the country of origin, thus avoiding a long detour between two areas of a country (Figure 13).

Figure 13. Northbound capitol corridor train in Union City.

Source: Image by Wikimedia Commons.

Cultural Tourism

In a narrow sense, special interest holidays (vacations) are essentially motivated by cultural interests, such as trips and visits to historical sites and monuments, museums, and galleries, artistic performances, and festivals, as well as lifestyles of communities. In a broad sense, including also activities with a cultural content as parts of trips and visits with a combination of pursuits.

D

Dacha

A Russian country cottage, commonly owned or part-owned through a cooperative, or rented by city dwellers for holidays (vacations).

Dam

A structure built across a river to hold back water for such purposes as flood control, irrigation, or storage. Sometimes a distinction is drawn between a dam and a barrage, the former but not the latter being used for power generation. Therefore, a well-known example is Hoover Dam on the border of Arizona and Nevada, USA, which is a major tourist/visitor attraction.

Damper

Australian expression denoting a bush loaf made from flour and water. It symbolizes outback tourism where travelers gather around a camp fire with tea boiled in a tin container called a Billy.

Datum Level/Line

The zero-altitude base for the measurement of elevation. For British official maps heights above the sea level are derived from the mean sea level at Newlyn, Cornwall, England.

Day Rate

The rate available in some hotels for daytime use of bedrooms, particularly common in airport hotels.

Day Visitor

A visitor who does not stay overnight in the country or place visited. Also known as an excursionist.

Day Visits Survey (DVS/UK)

Household sample survey conducted every second year since 1992 (pilot) and 1994 (full year) and sponsored by a consortium led by the Countryside Agency and the Department of Culture, Media, and Sport.

Daylight Saving Time

Local time observed by certain countries for part of the year, normally one hour in advance of the local standard time, in order to extend the period of daylight at the end of the working day; in the UK known as British summertime (BST).

Day Return Ticket

A ticket available at a lower price if you go and come back on the same day.

Days of Service

For most purposes the calendar week in travel and tourism begins on Monday. In timetables and other published material, days of service are often designated by numerals 1–7, beginning with Monday = 1.

Deadhead

- Aircraft or another vehicle traveling without passengers or cargo; e.g., deadhead flight.

- Airline or another carrier's employee traveling free (Figure 14).

Figure 14. Deadhead bus.

Source: Image by Wikimedia Commons.

Decorate

- To paint a room or a building or to put new wallpaper in a room;
- To cover something with pretty or colorful things to make it look attractive or to celebrate an occasion.

Deck Plan

A plan displayed in a ship, showing the various decks and what can be found on them.

Delay

The interval between the time at which something was planned to happen and the later time at which it actually happens.

Delhi Belly

Commonly used term for diarrhea (diarrhea) suffered when traveling abroad as a result of eating unwashed fruit or drinking contaminated water. Also called Montezuma's revenge.

Delta

A broadly triangular area at the mouth of a river is formed by deposits of solid material on the river bed, which build up faster than they are moved by tide or other currents.

Demand for Tourism

Normally defined in terms of the number of tourists. In this, a distinction may be drawn between actual (those currently participating in tourism) and potential (those who do not participate currently but may so do in the future). Moreover, for marketing purposes, both have to be backed by the necessary purchasing power for the demand to be effective.

Demi-Pension

Half-board hotel tariff, which includes room, breakfast, and one main meal per day, usually dinner.

Democratization

A process whereby opportunities are extended to wider strata of society, as has been the case with travel and tourism.

Demography

The statistical study of populations. Such characteristics as geographical distribution and age, sex, and household composition are known as demographics and are of particular importance in tourism planning, development, and marketing, as they provide the basis for market segmentation.

Demonstration Effect

Tendency for an individual or a group to imitate the behavior of another and to assimilate it as one's own. This process is often observed in tourism, especially in developing countries, when local residents are influenced by the behavior of tourists and assume their characteristics.

Denied Boarding Compensation

Payment by an airline to a passenger with a confirmed reservation for a specific flight not honored by the airline, commonly for such reasons as overbooking. Most airlines also reimburse passengers for reasonable expenses incurred by them as a result of not being able to travel as scheduled.

Departure Lounge

A room in an airport where passengers wait to get on their planes after going through passport control and baggage check (Figure 15).

Figure 15. Departure lounge at Pune airport.

Source: Image by Wikimedia Commons.

Departures

The part of an airport terminal that deals with passengers who are leaving.

Dependent Pass/Permit/Visa

Terms used for a permit to stay in a country are issued to the spouse of a non-national working in that country.

Depot

A building where the vehicles used by a transport service are kept.

Depressed Area

An area with high unemployment, low income per head, migration (especially of young people) out of the area, declining traditional industries, and an unattractive environment. A major need in such areas is to attract new economic activities, and in many, tourism is seen to have major potential.

Designated Carrier

An airline that is licensed to operate a service between two countries.

Destination

A place to which something is sent, or to which someone is going.

Detour

A journey away from the usual or planned route.

Determinants of Tourism

Factors determining the scale and patterns of participation in tourism. Demand determinants of private travel include such economic and social influences as the standard of living. Therefore, supply determinants include the availability, quality, and price of tourist attractions, facilities, and services, and their promotion. Hence, the volume of holidays (vacations) abroad is additionally strongly influenced by the size of the country of residence, its geographical location, and the relative costs of domestic and foreign holidays.

Dip Card

A type of magnetic key card which is pushed down into a slot and then pulled out again when the door unlocks (Figure 16).

Figure 16. Image showing dip card.

Source: Image by Wikimedia Commons.

Directions

Instructions telling somebody how to go somewhere.

Direct Flight

A flight which goes from one place to another, though with stops in between, as opposed to a nonstop flight.

Disembark

To leave a ship, aircraft, or another vehicle.

Dish of the Day

A special dish served on the day not appearing on the printed menu.

Distance Decay

A geographical term measuring the (generally exponential) decline in the level of interaction between phenomena with distance. An important tool in tourism and recreational planning, by relating the population of generating areas and the attractiveness of destinations, in order to permit predictions of likely volumes of tourist flows and recreational travel (which usually decrease as the distance increases).

Diversion

A change to a planned or normal route in order to avoid.

Dollar Area

An area of the world where the dollar is the main trading currency.

Dome Car

A railway carriage with a glass roof designed for sightseeing. Also called bubble car or observation car.

Domestic Escorted Tour (DET)

American term for an inclusive tour (IT) with a courier within one's own country.

Domestic Independent Tour (DIT)

American term for an unescorted trip itinerary within one's own country prepared by a travel agent for an independent traveler.

Domestic Same-Day

Visitor for statistical purposes, a domestic visitor who does not spend the night in a collective or private accommodation in the place visited [World Tourism Organization (WTO)]. Also known as a domestic day visitor or excursionist.

Domestic Tourism Expenditure

Defined for statistical purposes as expenditure received as a direct result of resident visitors traveling within their country of residence [World Tourism Organization (WTO)].

Domestic Tourism Monitor (DTM/Australia)

Annual survey of overnight and day trip travel behavior of Australians aged 14 years and over conducted since 1978 by AGB McNair for the Bureau of Tourism Research (BTR), in order to provide estimates of volume of domestic travel, periodic comparisons of tourism activity, and information about characteristics and behavior of domestic travelers.

Domestic Tourist

A tourist who is visiting another area of the same country where he or she lives (Figure 17). For statistical purposes, a domestic visitor whose visit is for at least one night and whose main purpose of visit may be classified under one of the following three groups: (I) leisure and holidays; (II) business and professional; (III) other tourism purposes [World Tourism Organization (WTO)].

Figure 17. A domestic tourist takes a pose in the gates of Bhaktapur Durbar Square.

Source: Image by Wikimedia Commons.

Domestic Travel Study (New Zealand)

Annual study by personal interviews of residents over 15 years of age as part of the McNair Omnibus Survey. Results are published in the form of separate regional reports, a demographic report, accommodation, and transport reports, origin/destination report, and a general report.

Domestic Travel/Tourism

Travel/tourism by residents of a country to other areas within the same country, as distinct from international travel/tourism. Resident aliens are treated as residents in travel and tourism statistics.

Domestic Traveler

Any person on a trip within his/her own country of residence (irrespective of the purpose of travel and means of transport used, even though he/she may be traveling on foot) [World Tourism Organization (WTO)].

Domestic Visitor

For statistical purposes, any person residing in a country, who travels to a place within the country, outside his/her usual environment for a period not exceeding 12 months and whose main purpose of visit is other than the exercise of an activity remunerated from within the place visited.

Dormitory Town/Village

A town or village from which many residents travel regularly to work in a nearby town or city. Some of them are former thriving resorts.

Double-Booking

The act of booking two people into the same hotel room or the same seat on a plane at the same time by mistake.

Double-Park

To park alongside a car which is already parked at the side of the street.

Down Time

An aviation term used in two different ways: (I) as the time period an aircraft is on the ground; (II) to indicate the time of landing.

Downgrade

To change to an inferior standard, as in moving an airline passenger or hotel guest or their reservations to an inferior seat or accommodation, whether initiated by the airline or hotel or by the passenger or guest.

Draught (Draft)

As a nautical concept, the depth of water which a vessel draws, i.e., the depth of the bottom of a ship below the water surface.

Drinking-Up Time

A period allowed in a public house after official closing time, when drinks already bought may be finished.

Drinks Trolley

A trolley on an aircraft, with various drinks that are served by stewards or stewardesses. Also called bar trolley (Figure 18).

Figure 18. Image showing drinks trolley.

Source: Image by Flickr.

Drive-In (Restaurant)

Type of establishment in which food is ordered by and served to motorists in their parked vehicles, to be found mainly in North America, and to be distinguished from drive-through (restaurant).

Dual Carriageway

A road with two lanes in each direction, with a barrier between the pairs of lanes.

Dude Ranch

A cattle-breeding establishment in North America providing tourist accommodation, riding expeditions and the opportunity to sample the cowboy lifestyle.

Dumb Waiter

- Waiter's sideboard in a restaurant;
- Small food lift between kitchen and dining room located on separate floors.

Dumping

Term used in international trade to describe selling for export at a significantly lower price than in the domestic market and often below the cost of production. The practice tends to be adopted for two main reasons.

One is the desire of the exporting country to penetrate a foreign market, another is to dispose of surpluses in order to avoid price reductions in the domestic market. Dumping is prohibited under the Treaty of Rome and by the World Trade Organization (WTO).

Dune

A ridge or hill of sand, deposited by the wind and characteristic of coasts and deserts where it is not stabilized by vegetation it can migrate downwind. It is therefore of significance for tourism and recreation because of their vulnerability to human pressure, which can cause severe erosion, particularly in coastal locations backing popular beaches or were used by off-road recreation(al) vehicles, such as dune buggies.

Duplex/Duplex Accommodation

Accommodation unit on two floors connected by a private stairway. When consisting of two rooms, one is normally used as a bedroom and the other as a living room, or both may be used as bed-sitting rooms.

Duplicate Reservation

Two or more reservations of a hotel room or passenger seat for the same person for the same night or journey when only one will be used. Also called double booking.

Duty-Free

Term applied to: (I) goods on which tax or customs duty is not levied; and (II) shops at international airports, ports, and ships in which passengers traveling abroad can buy such goods. Duty-free allowance denotes the quantity of dutiable goods allowed to be taken into a country without paying tax or duty. Duty-free allowances between European Union (EU) countries were abolished in 1999.

Dwell Time

In transport, the period of time a vehicle is at rest at a terminal.

E

Early Arrival

One who arrives at a hotel or another establishment before the date of reservation or before the arranged time.

Easement

A right, such as a right of way, that one owner of land has over the land of another.

Ecosystem

Ecological system, a system in which living organisms interact with each other and with the environment in which they live.

Ecotourism

Open to differences in interpretation but commonly denoting ecologically sustainable trips and visits to enjoy and appreciate nature, which promote conservation, have low visitor impact, and includes the involvement of local populations. It generally focuses on small-scale activities in well-defined areas, often under some designated form of protection, and on retention of the local traditional economy as a major employer (Figure 19).

Figure 19. Mangalajodi ecotourism.

Source: Image by Wikimedia Commons.

Ecotourist

A responsible tourist who eats local food, uses local transport, lives the lifestyle of local people, and respects their culture, and in general does nothing to damage the local environment.

Educational Trips/Visits

- Trips/visits undertaken primarily for education purposes, e.g., by school parties;
- Familiarization trips or 'fam trips'

Efficiency

American term for a hotel or motel room or apartment with cooking facilities.

Egocentrism

The perspective of one preoccupied with oneself and relatively insensitive to others, i.e., self-absorption, and self-centeredness.

Eighth Freedom

In civil aviation a right (not specified in the Chicago Convention) to carry traffic between two points within the territory of a foreign country, more commonly known as cabotage rights.

Elapsed Flying Time

The actual time spent in flight between two points, i.e., after allowing both for time on the ground and for any time changes.

Elapsed Travel Time

Of particular relevance in air travel, the actual time taken to travel between two points, after allowing for any time changes. In order to calculate the elapsed time, i.e., how long the journey actually takes, it is necessary to convert the local departure and arrival times into equivalent Greenwich mean time (GMT).

Electronic Funds Transfer (EFT)

Transfer of money between accounts through an electronic communication system, which provides direct links between terminals in retail outlets and computers in banks. Such systems are in growing use in hotels, restaurants, and travel agents.

Electronic Ticketing

First introduced by low-cost US domestic airlines in the early 1990s, electronic ticketing was extended to international flights by the mid-1990s. Commonly operated with cards issued by airlines, which enable passengers to book a flight by telephone or through travel agents and obtain a boarding pass from a machine at the airport, having paid with a credit card (Figure 20).

Figure 20. General ticketing machines in Singapore.

Source: Image by Wikimedia Commons.

Embargo

Restriction or prohibition by a government or another authority, originally of shipping and of goods entering or leaving a country. Now used in a wider sense, e.g., when referring to a ban on airline flights and on release of information.

Embark

To go on board a ship, aircraft, or other vehicles.

Employment Pass/Permit/Visa

Terms used for a permit to stay and work in a country issued to non-nationals, also known as labor permit or work permit and under other names in different countries.

Empty Leg

An empty flight between two consecutive scheduled stops.

Empty Nesters

An American term for couples whose children left home, when more time and money tend to be available to be spent on non-essentials, including tourism, and eating out.

En-Pension

Inclusive terms for hotel stays. En pension or full-board terms normally include room and three meals per day (breakfast, lunch, dinner), Demi-pension or half-board terms normally include room, breakfast, and one main meal, usually dinner.

Ensuite

Term used in such descriptions as 'bedroom with bathroom En suite' or 'bedroom with En suite bathroom,' i.e., with private bathroom attached.

Endemic

Adjective denoting that which is regularly found in certain places, such as an endemic disease.

Energy Management

Systems and processes used by companies and other organizations to reduce their use of electricity, fuel, and other energy resources for cost or environmental reasons. Such approaches are core components of cost control and environmental programs in travel, tourism, and hospitality operations.

Engel's Law

One of the generalizations put forward by German political philosopher Friedrich Engels, stating that as income increases, the proportion of it spent on food declines (and the proportion spent on consumer goods and services) increases.

English Breakfast

Breakfast is commonly served in the British Isles, normally consisting of juice and/or cereal, main dish, toast, butter, and preserves, tea, or coffee. The main dish usually consists of meat, fish or eggs or combinations of these. In Scotland, called Scottish breakfast.

English Service

Style of restaurant table service, mainly used for private functions, in which food is not portioned in the kitchen, but first presented to guests, as, e.g., whole joints of meat, before carving by the host or by staff.

Entrée

In menu terminology, originally a dish served before the joint of meat; nowadays, the main dish, which may be meat, fish, poultry or any other (Figure 21).

Figure 21. Entrée

Source: Image by Flickr.

Entry Requirements

Travel documentation required to enter a country, specified by its government, which may include a passport or another proof of citizenship, visa, and other documents such as vaccination certificates. Compared with international trade in goods, much international tourism is a relatively free market with a high degree of freedom of movement of people between countries. Moreover, this freedom is increasing, as border controls are reduced and simplified, but entry requirements of varying intensity continue to be applied by most countries to control immigration and for political or other reasons.

Entry Tax

Tax levied on passengers entering a country, usually foreign visitors, at airports but sometimes also at other entry points.

Environment

The surrounding conditions or influences. May refer to physical, e.g., natural or built environment or some other set of conditions or influences as, e.g., economic, social, cultural, or political environment.

Environmental Audit

A formal evaluation to assess the impact of a company's activities on the environment. A site audit focuses on a location such as a leisure complex; an operations audit examines the environmental impact and performance of a company's business processes.

Environmental Impact Assessment (EIA)

A set of procedures to assess in advance the likely environmental effects of a development project. Such an assessment is required by law in many countries for certain types of development, including the USA and the countries of the European Union (EU).

Environmental Management Systems (EMS)

Range of techniques used to monitor or manage the impacts of business and industry operations on the environment, such as environmental audit and environmental impact assessment (EIA), as well as carrying capacity assessment and cost-benefit analysis.

Equator

The longest parallel of latitude, equidistant between the north and south poles, where the sun is overhead at the spring and autumn equinoxes. Traditionally, passengers by sea crossing the equator for the first time undergo a ritual immersion by Neptune, God of the Sea.

Equinox

Time when sun is directly overhead at noon along the equator and day and night are equal length: vernal equinox about 21 March and autumnal equinox about 22 September.

Ergonomics

Systematic study of the relationship between man and his working environment with the application of anatomy, physiology, and psychology.

Erosion

The wearing away of the land surface by the action of running water, ice, and waves to produce a sequence of landforms. This is a slow and inexorable natural process, the rate of which depends on slope, climate, and rock composition and structure; thus, soft coasts of sand and clay are particularly vulnerable, with measurable retreat of the coast in historic times (Figure 22).

Figure 22. An example of sheet erosion shows how small rocks can protect the surface of the soil from the impact of raindrops.

Source: Image by Flickr.

Escarpment

Also shortened to scarp, a steep slope usually in areas of gently dipping sedimentary rocks where more resistant rocks overlie more easily eroded rock. It can give rise to striking scenery.

Escrow

An arrangement whereby payment for goods or services is held in a separate bank account or by a third party and released only when the goods or services have been supplied.

Esky

A portable insulated box-like container used in Australia to transport and keep food and drink cold for picnics, barbecues, and other outdoor functions.

Esplanade

A leveled raised ground along the seafront in a coastal resort, carrying a road and a promenade between the road and the seashore.

Establishment

A separate place of business at a separate address, in which one or more economic activities are carried on.

Ethnic Food

Food of a particular country, region, or racial group, such as Caribbean, Chinese, Indian etc.

Ethnic Tourism

Visits to places inhabited by indigenous and other exotic people, to observe their lifestyles and cultures, e.g., the Assam hill tribes in India, the Lapps in Northern Scandinavia, and the San Blas Indians in Panama.

Ethnocentrisms/Ethnocentricity

The belief that the attitudes and behavior of one's own ethnic group are superior to those of others, and the tendency to apply the standards of one's group to other groups. As tourism brings together people of different backgrounds and cultures and is therefore sometimes claimed to create a better appreciation of other people and their ways of life and to be a lever for change of attitudes and behavior on the part of residents as well as visitors.

Eurail Pass

First-class rail ticket for residents other than those of Europe and North Africa for unlimited travel within 17 countries for 15 days to 3 months of consecutive travel. Different prices apply to those under 12, 12–25, and over 25 years of age.

Eurail Select Pass

First-class rail ticket for residents other than those of Europe and North Africa for unlimited travel in three adjoining European countries for a maximum of 10 days. Different prices apply to those under 12, 12–25, and over 25 years of age.

Euro

A currency introduced in the European Union (EU) on 1 January 1999 and since adopted by 12 member countries (excluding Denmark, Sweden, and the UK). The twelve countries agreed to link their own currencies to the euro for three years to 31 December 2001, during which time the national currencies continued to be traded alongside the euro. Moreover, Euro notes and coins were introduced in January 2002. Thus, after a short transitional period, the euro became the only currency in which payments can be made (Figure 23).

Figure 23. Image showing euro coin.

Source: Image by Pixabay.

Euro Budget

Special return airfare available on scheduled flights in Europe as a promotional business class fare. Must be booked and paid for at the same time. A cancellation charge applies; the booking may be changed on payment of the difference between Euro budget and normal business class fares.

Euro Domino Pass

Unlimited travel rail ticket for those residents in Europe for at least six months available in first or standard class for three to eight days in a one-month period within one of 28 countries in Europe and North Africa. Different prices apply to those under 12, 12–25, and over 25 years of age.

Euro Lines

The name of a network of scheduled international motorcoach lines introduced by the International Road Transport Union in cooperation with 40 European coach operators.

Europa Bus

International network of motorcoach services of European railways providing inclusive tours (ITs) by luxury coaches with multilingual guides and using quality hotels.

Euro Pass

First-class ticket for residents other than those of Europe and North Africa for unlimited travel in France, Germany, Italy, Spain, and Switzerland for between 5 and 15 days within a two-month period. Different prices apply to those under 12, 12–25, and over 25 years of age.

European Prize for Tourism and the Environment

An award established by the European Commission in 1995, which aims to identify the perfect balance between tourism and the environment. The award is made to the destination that most innovatively and effectively develops a tourism policy that respects the environment.

European Year of Tourism 1990

The year marked as such under the auspices of the European Commission, aimed at heightening the awareness of the political, economic, and social importance of tourism. A total of 18 European countries were involved and more than 2000 related events were held in the UK alone.

Eurostar

Passenger rail service operating through Eurotunnel between London and Brussels, Lille, and Paris, jointly by Eurostar (UK) Ltd, French Railways and Belgian Railways under the commercial direction of Eurostar Group.

Eurostat

Statistical Office of the European Communities, which collects, analyzes, and publishes data on the European Union (EU) and its member states, including travel and tourism.

Excess Baggage

Amount of baggage (luggage) exceeding baggage allowance set by the carrier, which may be carried on payment of an additional charge. Under the weight system used by many airlines, the normal charge for every kilo of excess weight is 1% of the applicable one-way first-class fare; under the piece system, the excess charge depends on the distance flown, with additional charge for the actual weight and size.

Excess Mileage

Percentage supplement payable by an airline passenger if the maximum permitted mileage (MPM) between two points is exceeded. The additional amount is assessed at 5%, 10%, 15%, or 25%, corresponding broadly to the additional mileage.

Excess Value

Amount which may be declared by airline passengers at the time of check-in and by which, on payment of a fee to the airline, the possible compensation in the event of baggage loss or damage is increased. Excess value is applicable to both checked and unchecked baggage but is not an insurance: the airline will not necessarily compensate to the full amount and negligence has to be proved on its part.

Excursion

Generally, a short pleasure trip, also a sightseeing trip (usually optional) as part of an inclusive tour (IT) (Figure 24).

Figure 24. Vietnam river excursion.

Source: Image by Wikimedia Commons.

Excursion Fares

Special return airfares by airlines on scheduled flights, which are subject to various restrictions, such as specifying minimum and/or maximum length of stay, thus preventing dilution of full-fare revenue. They are less restrictive than APEX or PEX fares, e.g., allow stopovers if specified at the time of booking. Whilst more expensive than APEX and PEX, they are fully refundable.

Exile

A person is compelled by political or other circumstances to live away from his/her native country.

Expatriate

A person is living voluntarily away from his/her country of citizenship, not necessarily permanently but relatively long-term. In statistics of domestic and international travel/tourism, expatriates are treated as residents of the country in which they live.

Expressway

An American multi-lane divided highway designed for fast through traffic, with limited access and no surface crossroads, which may be a freeway or tollway.

Externalities

The impact of the actions of one individual or body on another, over which the affected individual or body has no direct control. The effects may be positive (i.e., beneficial) or negative (i.e., adverse). E.g., a new tourist/visitor attraction may create positive externalities for residents of the area by bringing additional spending power and employment but also negative externalities, such as increased traffic and congestion.

Extrapolation

Calculation from known terms, such as predicting a future situation on the basis of past data or events. For example, one approach to forecasting future visitor numbers to a destination is to project past trends into the future.

F

Facilitation

Action or process of moving forward, promoting, or rendering something easier. In travel and tourism, facilitation usually refers to measures taken by authorities to remove restrictions on movement, such as reduction and simplification of border controls, or marketing facilitation, as provided by tourist boards to support firms engaged in tourism through, e.g., organization of trade shows and workshops, familiarization trips, and provision of travel trade manuals.

Facsimile Transmission (Fax)

High-speed electronic transmission of an exact copy of a document from one location to another using the national or international telephone networks, which was the fastest-growing means of business communication in the 1980s.

Factory Tourism

Trips and visits by members of the public to working factories to observe processes, which may also enable them to purchase products from a factory shop.

Fall

Term used with the name of an outsized falls, like Niagara Falls in America or Victoria Falls in Africa.

Familiarization Trip

Trip ordinarily organized by tourer boards and similar organizations for journalists, tour operators and travel agents in generating areas to go to destinations and to become accustomed to their attractions, facilities, and services. These journeys, popularly referred to as fam journeys, could also be viewed as a part of commercial and together of the facilitation techniques utilized by business enterprise organizations to help the business enterprise business. Furthermore, they even have a promotional material role in providing opportunities for influencing communication and distribution channels for travel and tourer product and thru them ultimately potential customers.

Family Room Building

Room massive enough to accommodate a family, sometimes with four or a lot of beds.

Family-Style Service

Style of edifice tableware that during which bowls and platters of food are placed on the table from which guests facilitate themselves.

Far East (FE)

It describes conjointly however not exactly countries of east and south-east Asia, as well as China, Korea, Japan, and countries to the south of those. This can be the quickest growing destination region of the globe in international business enterprise with rates of growth well on top of the worldwide average. It is therefore conjointly AN more and more necessary business enterprise generating space.

Fare

Generic term for the value charged by a carrier and also the value of a price ticket for surpass transport.

Fare Construction Unit (FCU)

The basis of airfare calculations established by the International Air Transportation Association (IATA), and in use between 1974 and 1989, replaced by neutral unit of construction (NUC) on one Gregorian calendar month 1989.

Farm Business Enterprise/Farm Keep Tourism Vacation (Vacation)

Tourism centered on an operating farm, that takes numerous forms however invariably includes accommodation and infrequently alternative services for users. The two main forms are with repaired accommodation in or adjacent to the farm premises and with self-catering accommodation supported the farmland, e.g., cottages, caravans, and bivouacking sites. Moreover, farm business enterprise could also be seen as a type of rural business enterprise and is additionally known as agritourism, agricultural business enterprise, agrotourism, farm keep business enterprise and vacation farm business enterprise, particularly in North America.

Fast Food

Food ready, served, and consumed quickly on the premises or removed. Hamburgers, fish, and chips, and pizzas are typical examples (Figure 25).

Figure 25. Image showing fast food.

Source: Image by Flickr.

Fast Food Outlet

A restricted menu retail outlet giving fast, usually counter, service, and takeaway service. Sometimes supported systems with commonplace recipes, procedures, and product, several nutrition retailers are chain-operated with complete names.

Fast Lane

- The outer lane on an expressway;
- Slang term for prime living.

Fathom

A nautical measure of depth of water, based on the span of the extended human arms, corresponding to 6 feet or 1.829 meters.

Featherbedding

A apply of making extra jobs or spreading work by putting limits on production, requiring a lot of folks than necessary to try employment, or requiring the performance of superfluous work.

Federation

Designation unremarkably utilized by or applied to organizations with institutional instead of individual membership, within which every constituent member retains management over its own affairs. For example, a world organization could encompass national associations, like the Federation of Nordic Travel Agents' Associations.

Fee-Based Evaluation

An approach to travel agent's remuneration within which a mark-up is additional to the price of services provided per an agreement schedule.

Feeder Airline

A local airline in operation scheduled services between far fields and a significant airport.

Feeder Route

A local route of scheduled airline services between far fields and a significant airport.

Feeder Vessel

A small ship accustomed carry hundreds discharged at a port from larger vessels, to varied destinations.

Ferry

A boat (ferryboat) or a craft (air ferry) accustomed transport passenger, product, and/or vehicles on an everyday route between 2 points, sometimes across a stream, lake, or alternative body of water. Moreover, as a result of the coast of Europe is very indented, the ferry generally and also the automotive ferry specially are particularly outstanding options of European travel and business enterprise on the coast between the Baltic within the north and also the Aegean Seas within the south (Figure 26).

Figure 26. A ferry traveling across the Nile.

Source: Image by Flickr.

Ferry Mileage

The distance a ship, craft or another vehicle has got to travel while not passengers: (I) to come back to its base; (II) to deliver it from one location to another; (III) to maneuver it to and from a maintenance base.

Field Sports

A general term, particularly within the U.K., to hide fishing, looking, and shooting, conducted for the most part on or over non-public land, often in some type of multiple land use. Fishing and shooting, still as being undertaken by landowners and their friends, are let commercially to guests. Hence, fishing for fish, and shooting of grouse, pheasant, and red deer; fishing and shooting might also be available to those staying at hotels.

Fiords/Fjords

Heavily frozen coastal valleys that are invaded by the ocean, characteristically deep, however with a shallow threshold at the seaward finish. Therefore, they share several of the characteristics of frozen lakes and infrequently offer spectacular scenery, as within the western coasts of the Kingdom of Norway and European country, Canadian province, and South Island, New Island. They're necessary business enterprise resources, each for cruising and for those moving by land.

Flag of Convenience

The application of shipowners of registering their ships in countries apart from those of their own residence ports, for such reasons as avoiding taxes, government laws, high wage rates, and manning needs.

Flagship

Originally ship having an admiral on board; these days, the biggest or best ship of the fleet, whether or not that of the navy or of a company. The term is additionally accustomed to denote the leading member of a gaggle or chain of properties, once referring, e.g., to hotels or stores.

Flight Attendant

Member of cabin crew of a craft, conjointly referred to as air steward (ESS) or flight attendant.

Flight Coupon

The portion of a traveler price ticket that covers a flight between specific points, given at arrival time. Additionally, to 1 or a lot of flight coupons, the price ticket contains provision workplace and airline audit copies.

Flight Deck Crew

The pilots, flight engineers and navigators are liable for flying a craft, as distinct from cabin crew.

Flight Numbers

Flight designations unremarkably consisting of 2 letters (airline codes) and 3 figures. Some domestic and conjointly intra – European flight numbers have four figures.

Flight Recorder

Device carried within the craft that records technical info regarding the flight and that ought to be recoverable just in case of an accident. Conjointly referred to as in-flight recorder and said because the recording machine.

Floor

As a designation for a similar level of a building, the ground list differs within the United Kingdom (UK) and also the USA: within the UK the ground at street level is termed the bottom floor and within the USA the primary floor.

Flotilla Cruising

An arrangement within which a gaggle of yachts sails in an exceedingly fleet underneath the steering of a lead boat with a little crew sometimes consisting of a skipper in overall charge, AN engineer, and typically conjointly a hostess.

Fly Cruise

A holiday (vacation) arrangement consists of a scheduled or charter flight to and/or from a port and an ocean cruise (Figure 27).

Figure 27. Carnival cruise lines, fly-cruise.

Source: Image by Wikimedia Commons.

Fly Drive

An arrangement of scheduled or charter flight and automotive rent (rental), unremarkably engaged at a similar time or as a package at AN comprehensive value.

Fly Rail

An arrangement consisting of air and rail travel, unremarkably engaged at a similar time, or as a package. It is therefore going to embody a rail pass, that entitles the holder to unlimited rail travel at the destination for a nominal amount of your time.

Flying Boat

A craft that may land on and pop out from water, within which a ship is each body and float.

Foehn

A heat dry wind characteristic of mountainous regions, typically given its own name.

Food Court

A line of work thought within which a variety of various food retailers share a standard intake space. Sometimes comprising self-service nutrition retailers, food courts are usually situated in searching malls, transport terminals and alternative places with a high density of foot traffic. Therefore, conjointly major vacation (vacation) centers.

Footpaths

Narrow, unpaved routes for pedestrians across country in European country and Wales. Most are public rights of manner across farmland and thru woodlands. Moreover, native authorities have had to arrange definitive maps of such ways. In the main used for casual walking within the country by day guests.

Foreign Freelance Tour (FIT)

American term for AN unescorted trip itinerary abroad ready by a factor for the freelance traveler.

Foreign Escorted Tour (FET)

American term for AN comprehensive tour abroad with a messenger.

France Rail Pass

Unlimited travel rail price tag offered in 1st or commonplace category for 3 to 9 days at intervals a one-month amount within the French National Railways network. Totally different costs apply to those beneath twelve, 12–25, 26–60, and 60+ years older, still like a French Saver plan once 2 to eight individuals travel along.

Free Port

A port within which payload could also be unloaded and held on while not the imposition distinguished duties and taxes, that is of explicit price once foreign product are to be re-exported (Figure 28).

Figure 28. Freeport – Malta.

Source: Image by Wikimedia Commons.

Freedom of the Seas

A principle of jurisprudence that no state has any imposing right within the open seas and navigation on the open seas should be freed from any interference.

Free Sale

An arrangement that allows a tour operator to sell a united variety of edifice rooms while not 1st checking handiness with the edifice, whereby reservations is confirmed now to the shopper.

French Service

Style of edifice service, within which the food is portioned and organized on silver salvers within the kitchen; the salvers are placed within the edifice on a table with a tiny low heater, and therefore the food is served from the salvers on to the guest's plate.

Frequent User Programs

Promotional programs, conjointly referred to as loyalty schemes, usually targeted at business travelers and operated by carriers, edifice, and automotive rent still as alternative firms, within which the frequent (loyal) user of the company's services earns free or discounted travel, edifice stays or alternative edges.

Moreover, some programs are operated by one company, others collectively by 2 or a lot of firms, e.g., AN airline and an edifice.

Funicular (Railway)

A railway line created for the ascent of a mountain, within which the burden of AN ascending automotive is part or totally balanced by the burden of the raining automotive.

Fuselage

The main body of AN craft contains the traveler and payload compartments and therefore, the deck, i.e., excluding the wings, control surface, and framework.

Fifth Freedom

The right to use a carrier of 1 country to require passengers between 2 alternative countries.

Final Destination

A place reached at the tip of a journey once stopping at many places on the way. Conjointly referred to as final destination.

First-Class

Referring to the foremost high-ticket and most comfy variety of travel or edifice.

Five-Star Edifice

A very smart edifice, with luxurious rooms and better costs.

Flag Airline

The main national airline of a rustic, seen because the representative of the country abroad.

Flightseeing

The observe or business of transporting tourists to otherwise inaccessible geographical area areas by chopper, for viewing the areas by air or for unionized hikes.

Floor Larder

A small space serving one floor of an edifice, wherever the ground waiter prepares trays to require to guests' bedrooms and brings back dirty plates and glasses once use.

Fly-Drive Vacation

An arrangement wherever the traveler flies to AN airdrome and incorporates a rented automotive awaiting him or her to choose up, the rent of the automotive being paid beforehand as a part of the package value.

Fly-Stay

Referring to a meeting wherever you book an edifice at constant time as booking a ticket.

Fog Lights

Very bright red lights at the rear of an automotive, that are lit once driving in fog.

Free-Carving

A style of snowboarding that focuses on carving deep tracks within the snow with tight cornering instead of on doing stunts (Figure 29).

Figure 29. Snowboard carving.

Source: Image by Pixabay.

Free Climb

Mountain or mountaineering steer clear off aids like spikes and ladders, though sometimes with ropes and alternative safety instrumentation.

Freediving

The extreme sport of submersion into problem for as long as doable while not the help of element tanks.

Freeway

A major thruway wherever no toll charge is formed.

Frequent Flyer

Somebody UN agency flies usually with constant company, and then gets special treatment.

Fringe Edges

Benefits received by staff additionally to traditional remuneration, like free meals, medical insurance, pension arrangements.

Front of House

The operational areas and employees, e.g., all components of an edifice, court, or edifice in direct contact with the client, like foyers, bars, and eating rooms. Employees in these areas are vital to the customer's quality perception of the tourer product. That is, essentially experiential and involves abundant customer-supplier interaction.

Full-Time Equivalent (FTE)

Standardized unit found by changing part-time (PT) employment into full-time for accounting and applied math functions. Once knowledge is offered, this is often sometimes done on the premise of hours worked. However, a typical approximation is to convert numbers utilized into a full-time equivalent (FTE) persons utilized on the belief that PT staff work on the average 0.5 the hours of full-time staff, which the freelance work full-time.

Futurism

Term originally accustomed sit down with a movement in art and literature representing a radical departure from ancient strategies. These days used conjointly to explain AN outlook accenting contemplation of and designing for the long run.

G

Galileo

One of 2 main European laptop reservation systems (CRS) established in 1987 by a pool crystal rectifier by Alitalia, British Airways, KLM, and Swissair and together with conjointly Aer Lingus, Austrian Airlines, Olympic Airways, SABENA, and faucet, with US Covian Corporation provision initial software system and United Airlines being a partner within the system. In 1992 Galileo and Phoebus Apollo incorporate to make Galileo International as a worldwide distribution system.

Galileo International

A global laptop reservation system (CRS) shaped in 1992 by the merger of the Phoebus Apollo network within the USA and therefore, the Galileo network in Europe, owned in equal shares by 3 North Yankee carriers (United Airlines, USAir, and Air Canada) and European partners of the previous Galileo system, together with SAS, former founding father of Amadeus. Nevertheless, in North American nation, Gemini acts because the native distribution company for Galileo; within the USA, the United Mexican States, and Japan, Galileo uses the Phoebus Apollo brand.

Galley

Originally a ship's kitchen, but today includes an aircraft room for food preparation, as opposed to an in-flight kitchen or a terrestrial food preparation facility.

Gallon (Gal.)

For liquids, an English cubic measure of capacity, and for dry goods like maize, a dry measure. The imperial gallon is 27,714 cubic inches and equals 4.55 liters; the wine-gallon is 231 inches and equals 3.79 liters in the United States. A gallon is divided into four quarts or eight pints, and one bulk barrel contains 36 gallons.

Game Reserve

An area created and managed to conserve wildlife that can also be utilized for research and tourist control. In the United Kingdom (UK), there are no game reserves, but there are National Nature Reserves.

Gaming

Playing games of chance for money or for the sake of money Gambling is typically forbidden in public places (such as public houses, restaurants, and hotels) in the United Kingdom (UK), subject to specific exceptions, unless the premises are licensed for gaming under the law.

Gangway

- In an auditorium, a passage between rows of seats;
- Ramp for ship passengers to board and disembark;
- A stairwell with wheels that is used to load and unload planes.

Gateway

Because of its location and/or transportation ties, a city, or another significant point of access to a country or region. London, for example, is the main entrance to the United Kingdom (UK), and Copenhagen is the key gateway to Scandinavia.

Gazebo

A structure on a house's roof, such as a projecting window, balcony, or similar structure with expansive views, as well as one built on the ground.

Gemini Computer Reservation System (CRS)

Galileo International, which functions as the local distribution firm for Galileo in Canada, is owned by Air Canada, Canadian Airlines International, and Galileo International.

General Agreement on Tariffs and Trade (GATT)

On January 1, 1995, the World Trade Organization (WTO) took over as the legal and institutional foundation of the multilateral trading system, superseding an international agreement on eliminating trade restrictions between countries. In addition, the WTO oversees and implements multilateral agreements that encompass all internationally traded services.

General Household Survey (GHS)

Based on a sample of roughly 12,000 houses, a multipurpose continuous survey of all individuals in a household was intended to give information on a wide range of topics about private households in the United Kingdom (UK). Moreover, the survey's flexible structure enables it to fulfill special demands, and interviews conducted as a trailer to the survey have been employed in recent years for day visit surveys, for example (DVS). Hence, the Office for National Statistics publishes the GHS findings in reports (ONC).

General National Vocational Qualifications (GNVQS)

A framework of vocational qualifications established for young people aged 16 to 19 in England, Wales, and Northern Ireland as work-related education in schools and colleges and as a vocational alternative to academic qualifications. They are based on major occupational categories and provide both job preparation and a path to further education; fields covered include hospitality and catering, as well as leisure and tourism.

General Sales Agent (GSA)

Agents who are assigned by an airline or another principal to handle different aspects of promotion, reservations, ticketing, and inquiries for that airline or other principal in a specific territory are known as sole agents.

Generating Country

The country whose citizens visit to other countries for the purposes of international tourism statistics.

Gets a Computer Reservation System (CRS)

International Aeronautical Telecommunications Society operates this system, which is primarily geared to address the needs of smaller airlines.

Geyser

A geyser is a hot spring that produces intermittent columns of water. They can be found in volcanic areas such as Iceland, New Zealand, and the United States, where geysers are popular tourist attractions (Figure 30).

Figure 30. The great geyser.

Source: Image by Wikimedia Commons.

Ghetto

Historically, the area of a town or city in Europe where Jews dwelt. Nowadays, it is widely used to refer to a neighborhood where members of a minority dwell or where a large number of visitors congregate (tourist ghetto).

Ghost Town

A word used to describe a once-thriving town that has been abandoned by all or most of its residents. Many of these towns, which are mostly found in Australia and North America and were once mining villages, are now historical tourist/visitor attractions.

Ghosting

To distinguish between moonlighting and not declaring one's sole paid employment to the authorities for taxation purposes.

Gîte

Self-catering holiday (vacation) lodging in the countryside, which can be an apartment, cottage, or the entirety or a portion of a house. Farm outbuildings, flats in old presbyteries or schoolhouses, and château wings are among the many that have been converted. For example, in France, there are approximately 10,000 gîtes. The titles Gîte de France and Gîte Ruraux de France signify that the lodging satisfies the organizations' requirements.

Giveaway

Customers and/or industry contacts are given promotional gifts such as branded travel bags, T-shirts, stationery, and souvenirs. Also referred to as promotional products.

Glacier

A massive mass of ice formed by the compacting and crystallization of snow and sliding downslope slowly under gravity's influence. There is a difference between piedmont and valley glaciers, with the former spreading out over a broad front and the latter being more tightly restricted and moving faster. Moreover, glaciers can now be found primarily in high and middle latitude mountainous areas, such as Scandinavia, the Alps, the northern Rockies, and New Zealand's South Island, where they have left their mark on the rugged scenery, over deepened valleys, and fjords of mountain country, making such areas appealing to visitors.

Glen

As opposed to strath, a deep narrow valley usually forms the route of a stream in Gaelic.

Global Distribution System (GDS)

Galileo International, commonly known as the global reservation system (GRS), is a worldwide computer reservation system (CRS) that can distribute airline and other travel services all over the world.

Global Tourism Regions

Africa, the Americas, East Asia and the Pacific, Europe, South Asia, and the Middle East (ME) are the six Regional Commissions of the World Tourism Organization (WTO) that are utilized for statistics purposes.

Global Warming

Global warming is generated by the release and accumulation of specific gases in the atmosphere, which allows solar energy to pass through but block heat from exiting. The greenhouse effect, often known as global warming, may result in rising sea levels, which might ruin low-lying coastal resorts, as well as ski resorts, coral reefs, and other tourists/visitor attractions.

Globalization

Globalization of markets and international enterprises to serve them, as well as a convergence of global tastes, product preferences, and lifestyles, have led to increased uniformity and market homogenization. Hence, travel and tourism stimulate cultural globalization while also displaying many of its characteristics.

Golden Age Passport

The card, which is available to persons aged 62 and up, grants unrestricted entrance to US National Park Service parks and sites.

Golden Era of Bus and Coach Travel

Term used in the United Kingdom (UK) during the interwar years of 1918–1938, when more passengers were transported by buses and coaches than at any time before or after.

Golden Week

Term used in Japan during a week in May when many public holidays fall on the same day, resulting in high demand for tourism facilities and services both in and outside of Japan, and when tour, transportation, and accommodation prices spike in reaction to the excessive peaking of demand.

Gondola

A light flat-bottomed boat with a high point at either end and one oar at the stern that is used to transport people and goods through Venetian canals (Figure 31).

Figure 31. Gondola at the Venetian.

Source: Image by Wikimedia Commons.

Gothic

The essential elements of this architectural style from the twelfth to sixteenth centuries in Europe are pointed arches and ribbed vaulting. The phrase is often used to describe period painting and sculpture.

Grand Tour

Between the late sixteenth and early nineteenth centuries, this term was frequently used to characterize travel by younger members of the English society to countries in Continental Europe. Moreover, undertook for what may now be referred to as wanderlust impulses to become 'learned and civilized' by exposure to European art, manners, and culture; much of it went place with tutors or servants, often over the course of two or three years. Hence, The Grand Tour, together with early travel to spas and coastal resorts, reflects the roots of modern tourism.

Gratuity

A gratuity paid willingly by a consumer to a business that provides a personal service, such as catering, hairdressing, or taxi transportation.

Gravity Model

One of several attempts to describe the elements that affect tourist flows and determine the volume of flows between places, based on Newton's law of gravitation. The 'mass' of the regions (e.g., population) and the distance between them (in linear, time, or cost terms) are the two primary factors: the larger 'the mass' and the smaller the distance, the greater the flows.

Great Britain (GB)

England, Wales, and Scotland are sometimes incorrectly referred to as Britain, which is also used to refer to the British Isles and the United Kingdom (UK).

Green Belt

An area of the countryside surrounding a large built-up area, which may include farmland, woodland, and other lands, in which building development is strictly controlled in order to prevent unplanned urban sprawl and the merging of neighboring towns, as well as to preserve the unique character of historic towns and to aid in urban regeneration.

Green Field Site

An undeveloped tract of land, usually in a rural region, that is being developed or is being developed.

Green Globe

The World Travel and Tourism Council has designed a global environmental management and awareness program for travel and tourism enterprises committed to improving their environmental practices (WTTC).

Green Holidays (Vacations)

Holidays (vacations) spent in the countryside doing environmental activities, usually supported by government and non-profit organizations.

Greenwich Mean/Standard Time (GMT/GST)

The local time at Greenwich, England's zero meridians, and the British Isles' standard time, from which the world's standard time is determined.

Greeter

A male employee who greets and directs visitors to their tables in a restaurant in the United States. A hostess is a female employee who works at a restaurant.

Grid Route System

A system of airline routes based on a number of airports from which flights are scheduled to operate on multiple integrated routes (Figure 32).

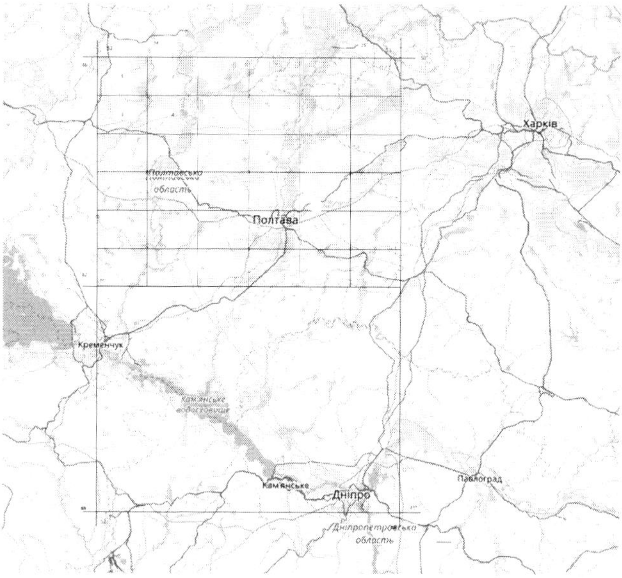

Figure 32. Russian grid system.

Source: Image by Wikipedia.

Grockles

In the United Kingdom (UK), a derogatory nickname for visitors. It is said to have originated in Torbay, Devon, and refers to tourists who resemble miniature Grocks (Grock was an internationally famous circus clown whose typical prop was a suitcase).

Gross Domestic Product (GDP)

In value-added words, a measure of the value of goods and services created by an economy over a period of time, usually a year or a quarter, excluding the value of intermediate items. The ratio of tourism receipts to GDP is frequently used as a measure of how important tourism is to a country's economy.

Ground Arrangements

A person or organization known as a ground operator, ground handling agency, incoming tour operator, or destination management company provides local services for tourists at destinations such as hotel transfers, automobile hiring (rental), and sightseeing. It is therefore known as land agreements in the United States.

Group Inclusive Tour (GIT)

An all-inclusive tour (IT) for members of an organized group who qualify for a scheduled airline's group-based air rate; the participants must travel together on the same outbound and return flights.

Grub

Colloquial term for food. Pub grub, for example, is cuisine offered in pubs such as sandwiches, pies, salads, and fast food.

Guaranteed Payment Reservation

Hotel reservations made in advance with payment assured even if the guest does not stay.

Gueridon

In a restaurant, a trolley from which the final preparation and service of a meal is performed. As a result, gueridon service was born.

Guest House

Also known in Continental Europe as a pension, it is a small, owner-managed establishment that provides solely lodging, food, and drink to its tenants. It is also known as a pension in Continental Europe. (b) A tiny state-owned hotel for official visitors in the far east (FE).

Guest Worker

A person who relocates to another country for the sake of employment. Temporary migration usually occurs from a country with heavy unemployment to one with labor shortages, as was the situation in the United States.

Guest Ledger

A list of all money owing to a hotel by a visitor that must be paid before the person checks out, such as lodging, bar bills, and phone charges.

Guidebook

A tourist guidebook that explains what to see in a location, where to stay, how to get around, and where to eat.

Gulet

A classic Turkish wooden yacht with a double bedroom and a big sundeck that can accommodate up to 15 people.

H

Hague Declaration on Tourism

The Inter-parliamentary Conference on Tourism, conducted jointly by the Inter-Parliamentary Union (IPU) and the World Tourism Organization (WTO), adopted a declaration in April 1989 in The Hague, Netherlands (WTO).

The Declaration contains ten tourist development principles and highlights the importance of a global approach and international cooperation involving both governments and the corporate sector.

Habitat

An ecological phrase that describes the physical circumstances that plants and animals require and utilize. One of the main goals of nature reserves is to conserve examples of all of the country's major habitats and species.

Hajj Pilgrimage

To Mecca, which all Muslims are required to visit at least once in their lives, and which takes place in the final month of the Moslem calendar (which varies from year to year) and is a notable example of religious travel/tourism (Figure 33).

Figure 33. The Kaabah (Kaaba) during Hajj.

Source: Image by Wikimedia Commons.

Handbill

A printed sheet used to advertise a variety of products and services, such as restaurants, and given to potential consumers or mailed via their letterboxes.

Hansom (Cab)

A two-wheeled one-horse carriage for two passengers with a driver mounted behind the reins that go over the roof, which is now primarily used for sightseeing rather than transportation between two sites.

Happy Hour

A promotional technique used to entice clients, designating a time when hotels, restaurants, and other places provide discounted drinks. Usually in the early evening.

Hard Currency

A currency with a high demand relative to supply and a constant or growing exchange rate is characterized.

Hard Tourism

Term used to describe large-scale tourism and indiscriminate tourist development motivated by commercial motives while ignoring its social, cultural, and environmental repercussions, as opposed to soft or alternative tourism.

Hawker Center

A group of privately run food stalls with shared tables and seats, often in an alfresco or temporary semi-outdoor setting, synonymous with inexpensive uncomplicated local food. Food is ordered from a variety of hawkers, who deliver food to consumers who may or may not be sharing tables. Popular in Southeast Asia and viewed as a must-do for travelers.

Head Wind

Wind blowing in the direction of a ship or aircraft, slowing it down.

Health Club

People who want to improve their health by exercising frequent a club or space in a hotel or apartment building with workout equipment and, in certain cases, a swimming pool.

Health Declaration Form

Some countries require visitors to fill out a form stating excellent health, disease-free status, and no recent contact with specific locations where certain diseases are endemic. As public awareness of the spread of AIDS among tourists grows, some authorities are requesting that health forms include an HIV provision. The OAG Guide to International Travel and other travel reference manuals have specific criteria.

Health Farm

People come to a residential clinic in the countryside to improve their health and beauty through diets, exercise, and relaxation.

Health Food

Natural meals or foods without additives, such as cereals, fruits, and vegetables, and yogurt, are said to be helpful to one's health.

Health Resort

A health-giving resort has specific facilities, such as air, sun, and mineral waters, that is visited for treatment, convalescence, and relaxation.

Health Tourism

Trips and visits to health resorts and other destinations whose major objective is health treatment, ranging from therapeutic treatments for various disorders to fitness and relaxation programs. Therefore known as health-care tourism. Many hotels and cruise lines, as well as places like health farms, provide some of these services.

Heart of England

The Heart of England Tourist Board encompasses the counties of Derbyshire, Gloucestershire, Herefordshire, Leicestershire, Lincolnshire, Northamptonshire, Rutland, Shropshire, Staffordshire, Warwickshire, Worcestershire, and some areas, and is one of ten English Regional Tourist Boards (RTBs).

Helsinki Accord

The European Conference on Security and Cooperation in Helsinki, Finland, adopted a statement of principles in 1975. In addition to recognizing the importance of international tourism, the 35 participating countries reaffirmed their support for administrative simplification and harmonization in the sector of international transportation. As well as their determination to encourage expanded tourism in a variety of ways.

Hemisphere

A plane traveling through the globe's center divides the world into two halves. The equator divides the earth's surface into northern and southern hemispheres, and the meridians 20° West and 160° East split the earth's surface into eastern and western hemispheres (Figure 34).

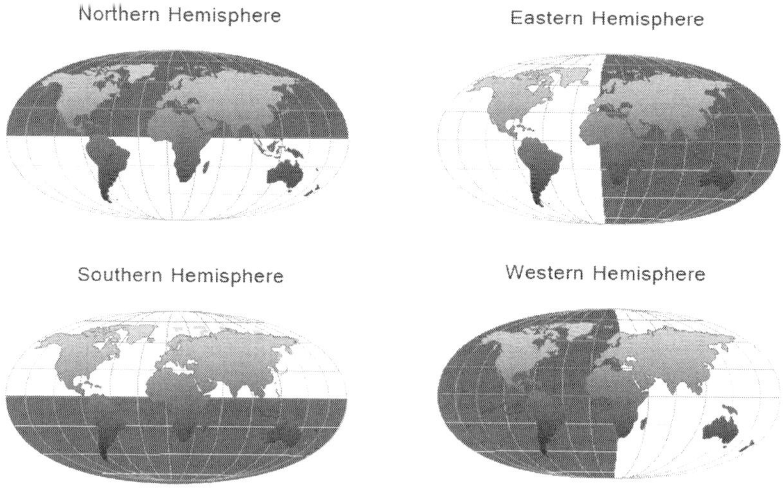

Figure 34. Hemispheres.

Source: Image by Flickr.

Hepatitis A (Infectious Hepatitis)

A liver infection that causes fever, stomach pain, and maybe jaundice. It is primarily transferred by contaminated food and water, but it can also be carried through feces. The risk exists all around the world, particularly in hotter climates where sanitation is lacking. Moreover, food and water safety, as well as personal hygiene, are the most important safeguards; short-term immunity is provided by immunoglobulin or another vaccine.

Hepatitis B (Serum Hepatitis)

Hepatitis B is a liver infection with symptoms similar to hepatitis A, but it is more infrequent and serious, and it can develop into chronic liver disease. Infected blood or body fluids, such as infected equipment or sexual contact, are the most prevalent ways for it to spread. This happens all across the world, although it's more common in impoverished countries with insufficient sanitation. Hence, the best way to prevent infection is to avoid exposure; a vaccine is also available.

Hepatitis C

A liver infection spreads in the same way that AIDS and hepatitis B do. There is no vaccination available. The easiest method to avoid infection is to follow the AIDS and hepatitis B prevention guidelines.

Heritage

Those parts of the environment, consisting of natural and man-made resources of exceptional value and importance that should be preserved for future generations.

Heritage Coasts

In England and Wales, government countryside organizations and local councils collaborate to define undeveloped coasts of special visual appeal. By the mid-2000s, 45 coasts totaling 1,540 kilometers (960 miles) have been delineated. Other plans can be found under rural conservation designation initiatives.

Heritage Tourism Holiday (Vacation)

Heritage tourism holiday is a type of special interest tourism that focuses on visits to historical sites.

High Tea

A large meal served late in the afternoon in the North of England, Scotland, and Wales, usually comprising of a cold or hot main dish, cakes, and tea.

Hiking

Walking for pleasure, especially in open and natural areas, and encompassing both day visits and extended holidays. Bushwalking, hill walking, and trekking are examples of activities that provide both a physical challenge and a way to see a lot of beautiful scenery, depending on the terrain and vegetation. Hitch-hiking is a term used to describe traveling by requesting rides from vehicle and truck drivers.

Hill Station

A high-altitude settlement commonly established by Europeans in former colonial areas as a respite from the summer heat and humidity of the lowlands, such as the British in India and the Dutch in Indonesia (Figure 35).

Figure 35. Nelliyampathy hill station.

Source: Image by Flickr.

Hinduism

Formulated between 500 BC and 500 AD as sacred texts in Sanskrit by the Brahmans, from which Hinduism developed Hindus' Asian polytheistic religion and social system with beliefs in the veneration of the cow and transmigration of the soul.

Hippie

Term used to describe people who reject middle-class norms and customs and, as a result, live communal and/or nomadic lifestyles.

Hold

Originally a nautical term, it today refers to luggage or cargo/freight storage space beneath an airplane or ship's deck.

Holiday

a) A day when labor is optional and rewarded by law or custom. (b) A vacation or time away from home outside of the United States.

Holiday Leave Loading

Payment of a 17.5% premium over an employee's standard income during a four-week annual holiday (vacation) in Australia. The loading was originally implemented to augment otherwise reduced income for workers who rely on overtime and to provide them with the resources to go on vacation, but it has now become a right for most Australians.

Holiday (Vacation) Frequency

The ratio of total holiday (vacation) trips to the number of people who take at least one trip in a certain time period (typically a year), i.e., the average number of trips taken by those who take any travels.

Holiday (Vacation) Propensity, Gross

In relation to the entire population, the total number of holiday (vacation) travels taken in a certain period (typically 12 months). This statistic, unlike net propensity, indicates the number of trips per 100 people and reflects the tendency for people to make many journeys in a given period.

Holiday (Vacation) Propensity, Net

The percentage of the population or a subset of the population (e.g., income group or age group) who take at least one vacation (vacation) excursion in a given year (usually 12 months).

This is a basic indicator of tourist participation obtained from national vacation (vacation)/travel/tourism surveys. Germany, Sweden, and Switzerland (70% or more), Denmark, the Netherlands, and the United Kingdom (UK; 60% or more) have the highest propensities in Europe (60% or more).

Holistic

To describe a system of interconnected components or a whole made up of orderly groupings of individual units. 'The constituents of the system include tourists, generating regions, transit routes, destination regions, and a tourist business,' writes Australian author Neil Leiper. Hence, these pieces are connected both spatially and functionally.

Honeypots

Strongly appealing locations that draw huge crowds and can become congested during peak hours. Honeypots can emerge spontaneously or as a result of deliberate policy, with the goal of offering tourist amenities and/or diverting visitors away from sensitive regions.

Hors-D'oeuvre

Pâté, hard-boiled eggs with mayonnaise, salad, or multiple foods presented together as mixed hors-d'oeuvres are examples of cold dishes served at the start of a meal in French (Figure 36).

Figure 36. Hors D'oeuvre by Auriole Potter.

Source: Image by Wikimedia Commons.

Hospitality Assured (HA)

The Hotel and Catering International Management Association (HCIMA) launched a British accreditation program in June 1998 in order to encourage improved standards of customer service as well as higher standards of professional organization in the hospitality sector.

Host Bar

Open bar is an American term for a bar supplied at private occasions where guests are allowed to free drinks.

Host Community

The local inhabitants of tourism destinations. Tourists may have a negative or positive impact on their culture, environment, and values; as a result, planners, and developers are becoming more aware of the importance of involving host communities in the planning and development process (community tourism).

Hostel

A place that offers cheap lodging and sometimes food to specialized groups rather than the general public, such as employees of a company, students, and young travelers (youth hostel).

Hostelry

A term for an inn that dates back to the Middle Ages and is currently most commonly used to designate a traditional inn.

Hostess

A woman who looks after passengers or guests (for example, an air hostess or a ground hostess) (e.g., in the USA. An employee who receives guests in a restaurant and shows them to their tables).

Hot Springs

Continuous flow of water in areas of current or recent volcanic activity, at temperatures ranging from 20°C to 100°C (65–212°F), in contrast to the occasional and powerful ejection of water from geysers. Iceland and New Zealand's North Island, for example, have a lot of them. Therefore, thermal springs are another name for them. They can also be found in non-volcanic places, such as Bath, England.

Hotel Establishment

Providing lodging, food, and drink to travelers and temporary residents as a reward, as well as meals and refreshments to other users, and frequently other facilities and services.

Hotel Classification

The classification of hotel accommodations according to their type and range of available amenities and services, which may be extended to other types of tourist lodging. Moreover, classification schemes, which are usually controlled by tourist boards, motoring organizations, and other similar organizations, sometimes include minimum standards for each kind or class of lodging and can be paired with hotel rating.

Hotel and Catering Industry/Services

Firms and establishments that provide paid lodging, food, and drink away from home, with differing definitions for specific reasons in different nations.

Hotel Garni

A hotel with minimal amenities and services, mostly confined to the provision of lodging with breakfast, but occasionally also with limited service of drinks, snacks, and light refreshments to residents, which is particularly common on the Continent of Europe.

Hotel Grading

Quality control of hotel amenities and services, which might be used for various types of tourist accommodations. Physical features, food, and drink, and/or other services may all be evaluated, and the results may be stated individually for each or collectively for the entire facility using letters, numbers, or symbols.

Hotel Industry

The term is mostly used to refer to hotels and motels, although it is occasionally used to refer to guest homes and boarding houses as well (Figure 37).

Figure 37. Le Méridien hotel in Thimphu, Bhutan.

Source: Image by Wikimedia Commons.

Hotel Registration

Guests are required to register when they arrive at hotels. (b) Hotel listing, resulting in an inventory that can be expanded to include other tourist accommodations. In order to be comprehensive, registration must be backed by legal power and administered by a government or statutory organization, which may impose certain minimum requirements. Moreover, in most countries, registration schemes are used for planning purposes and/or as a basis for hotel classification and grading under fire prevention or other health and safety consumer protection regulations.

Hotel Tariff

Price list or schedule for hotel services. Accommodation and related charges may be quoted "per room" or "per person," as separate charges or in more or less inclusive terms, with or without service and taxes. Therefore, different prices may apply to individuals and to groups, during the week and at weekends, and at different times of the year.

Hotel Representative

A person or organization hired by hotels to assist travel agencies and individuals with reservations. Such arrangements are particularly beneficial to independent hotels, as well as hotel companies, who would otherwise have to run their own reservation offices abroad.

Hotel Tax

Tax placed on staying guests by the central or local government or another agency and collected at the place of stay as a means of producing money; the earnings are sometimes used for tourism objectives. It is also known as a bed tax, a room tax, or a visitor tax.

Hotel Voucher

A coupon is a document that is frequently issued by a tour operator or travel agent to cover prepaid parts of a trip such as lodging and meals, and it is relinquished by the passenger upon arrival at the hotel, which then demands payment from whoever issued it.

Hotels and Similar Establishments

Hotels, motels, inns, guest houses, and boarding houses are all examples of this term used by tourism organizations. According to the World Tourism Organization (WTO), hotels, and comparable businesses are defined as being arranged in rooms with a minimum number of rooms; being managed by a single entity; and providing certain services such as room service, daily bedmaking, and sanitary facility cleaning.

Hub and Spoke System

A network of airline routes that runs via a major airport (hub) and serves as a hub for feeder flights to and from smaller airports (the spokes). This method, which also applies to other forms of transportation, is more conducive to larger load factors than providing direct connections between a large number of airports. Moreover, one of the main results of airline deregulation in the United States and the liberalization of international competition is its considerably increased occurrence in air transportation.

Hypothetical Fare Construction Points

Also known as fictitious construction points, this airline term refers to points (i.e., airports) placed in an itinerary in order to generate a reduced airfare. The customer does not fly to these locations; they are listed in the fare construction section of the airline ticket, but there is no flight voucher for them.

I

IATA Bonding Scheme

The International Air Transport Association (IATA) operates a bonding system to safeguard IATA airlines and ensure that they get paid in the event of a travel agent collapse. Scheduled flights with IATA airlines booked through IATA designated agents are covered by the plan. Other schemes can be found under bonding schemes (UK).

Iberia

Since the 1960s, the peninsula shared by Spain and Portugal in south-west Europe has been a popular holiday (vacation) destination for northern Europeans. Spain was one of the first countries to enter the mass inclusive tour (IT) market, and tourism has played a significant role in the country's growth.

Immunization for Travelers

Diphtheria, poliomyelitis, and tetanus, if not previously immunized. Poliomyelitis (b) poliomyelitis (c) poliomyelitis (d) poliomyelitis (e) poliomyelitis (e) poliomyelitis (e) poliomye (booster dose if immunized more than 10 years ago) (c) For places where hygiene and sanitation standards are less than optimum.

Implant Travel

As opposed to outplant, dedicated workers dealing with the account on the travel agency premises, a business travel team located in a large enterprise or organization makes travel arrangements for its employees.

Import Content of Tourism Expenditure

Imports of goods and services from other countries to meet the demands of tourists, which are then purchased by tourists. Therefore, import content covers goods and services purchased from other regions of the country at the subnational (regional and local) level.

Import Substitution

The substitution of domestically produced items for imported goods. The phrase is most commonly used in the context of the balance of payments to refer to goods rather than services. It is, however, equally relevant to services, with the same beneficial balance of payments effect when, for example: domestic holidays (vacations) are substituted for holidays (vacations) abroad.

Inaugural

In travel and tourism, the term "inaugural trip" or "inaugural flight" refers to the formal ceremonial debut of a new ship, aircraft type, or route (Figure 38).

Figure 38. Inaugural flights Thai LionAir.

Source: Image by Wikimedia Commons.

Inbound Travel/Tourism

Travel/tourism to a given country by residents of other countries.

Incentive Commission

A bonus or incentive paid by a principal (such as an airline or tour operator) to a travel agency in exchange for high sales.

Incentive Travel/Tourism

Travel paid for by a company as a reward for meeting sales or other targets, for great performance, or as a stimulus to future achievement by workers, dealers, or agents, sometimes with spouses. Many airlines provide discounted flights known as incentive rates for incentive groups, and specialist businesses known as incentive companies exist to design incentive travel programs for clients.

Inclusive Tour (IT)

A holiday (vacation) package that includes transportation, lodging, and sometimes other travel services such as sightseeing, and is sold at an all-inclusive price that is usually substantially lower than the cost of reserving the individual pieces separately. Moreover, although air-based inclusive vacations are the most visible, inclusive tours (ITs) can be built using any mode of transportation and any type of accommodation, as well as a variety of other services.

Inclusive Tour Fare (ITX)

Some airlines offer a tour-basing rate to tour operators who produce a certain amount of promotional literature for inclusive packages.

Incoming Tour Operator

According to the scope of its functions and services, which may include, for example, arrangements for conferences and exhibitions, a person or organization that provides local services for tourists at destinations, such as hotel transfers, car hire (rental), and sightseeing, is also known as an inbound tour operator, ground operator, ground handling agent, and destination management company.

Index Numbers

Summary numbers that represent relative changes over time with relation to a reference point, usually 100. As a result, an index number at a given time reflects the relative value at that moment in comparison to the value at the base time.

Indian Subcontinent

South Asia is made up of Afghanistan, Bangladesh, Bhutan, India, Nepal, Pakistan, and, in most cases, Sri Lanka. An area with low international tourism, attracting less than 1% of global visitors and receipts, with India accounting for more than half of the total.

Indian Summer

A period of calm, dry, moderate weather with clear skies but a hazy atmosphere that occurs in late autumn in the UK and the United States.

Indigenous Tourism

It is a word used to characterize tourist activities that incorporate indigenous people, i.e., people in their natural environments, and in which indigenous culture is a primary attraction.

Indirect Route Principle

The general guideline that allows passengers paying full rates on IATA airlines to divert from the direct route between any two points without incurring additional costs as long as the route's maximum allowable distance (MPM) is not exceeded, as defined in the tariff manuals.

Industrial Awards (Australia)

In Australia, there is a centralized wage and benefit system in place. The 'award wage' is the minimum rate for a certain activity or function. Enterprise agreements are being reached to cover terms and conditions at the enterprise level, including some in the travel, tourist, and hospitality sectors, resulting in decentralization of the existing system. Moreover, in travel, tourist, and hospitality firms, enterprise agreements may include the payment of rates above the award wage for increased productivity, as well as the waiver of 'penalty rates' for work done outside of 'typical' hours.

Industrial Heritage Attractions

Tourist/visitor attractions with exhibits of products and sometimes operating processes based on old industrial establishments, such as coal mines or shipbuilding yards.

Industrial Revolution

The transition from an agrarian and handicraft economy to an industrial and manufacturing economy, which began in England in the eighteenth century and extended around the world beginning in the early nineteenth century. Hence, population growth and urbanization, societal structural changes, advances in the standard of life, and transportation developments were all major aspects of the change of significance to travel and tourism.

Industrial Tourism

Trips and visits to places of employment to watch processes, such as breweries and distilleries, food processing and textile industries, as well as mines and power plants (Figure 39).

Figure 39. Place on industrial tourism path in Lafarge cement plant Trbovlje.

Source: Image by Wikimedia Commons.

Industrialization

The transition from an agricultural and handicraft economy to a manufacturing economy, as evidenced by job transfers from the former to the latter and internal migration from rural to urban regions, both of which are critical for tourism development.

Industry

In practice, certain economic activities are frequently carried out by a group of enterprises and facilities with a common interest, and these activities are referred to as an industry. The bond may be one of the types of products, use of materials or process, or a still looser one – such as their general function and place in the total economic activity.

Infini

With cooperation agreements with Abacus and Worldspan, a computer reservation system (CRS) owned by All Nippon Airways.

In-Flight

Activities during the flight while onboard an aircraft such as, e.g., in-flight catering, in-flight entertainment, in-flight magazines. These passenger services may be provided free or in some cases for payment.

Infrastructure

All types of construction required by a populated area in touch with the outside world in order to support and facilitate economic development. Roads and railways, harbors, and airports, as well as public utility services such as water supply, drainage, and sewage disposal, power supply, and telecommunications, all fall under this category.

Inland Waterways

Rivers and canals that can be navigated in nations (mostly in Europe) that have significant canal networks, which were originally intended for commercial traffic but are now widely used for water-based recreation and sailing vacations (vacations). Moreover, in the United Kingdom (UK), cruise waterways account for more than half of the 3,200 km (2,000 miles) network managed by the British Waterways Board, which is being developed for boating, fishing, and other recreational activities.

Inn

Unlike in the past, when the term was used to distinguish places that provided overnight lodging from taverns and alehouses. It now has no specific definition. It is therefore used indiscriminately for hotels and eating establishments that do not offer overnight accommodations. Although it is certainly more popular in smaller enterprises, it is not limited to them, as demonstrated by corporations like Holiday Inns.

Inner City

An area within a large city that is often plagued by economic and social issues. In many nations, tourism has recently been considered as a way to revitalize inner cities by attracting tourism-related enterprises and activities, as well as adding spending power to the area.

Innkeeper's Lien

Hotels have the legal power to keep guests' belongings in certain instances due to unpaid debts. Moreover, in most English-speaking nations, the right has been given statutory power based on English common law, which usually includes the right to sell the items held on lien.

INSPASS (Immigration and Naturalization Service Passenger Accelerated Service System)

An acronym for Immigration and Naturalization Service Passenger Accelerated Service System, an experimental automated immigration lane project that began in 1993 at J.F. Kennedy International Airport in New York and is open to US and Canadian citizens as well as nationals of countries participating in the US visa waiver program.

Integrated Resort

In general, a resort development in which all components, including as infrastructure, superstructure, and supporting services, are planned, constructed, and operated in a complementary and integrated manner. In contrast to ad hoc resort development, where specific features are created and operated independently, this word is utilized. In several nations, there is resort-specific legislation (Figure 40).

Figure 40. Marina Bay Sands integrated resort, Marina Bay.

Source: Image by Flickr.

Inter-Governmental Organization (IGO)

International organization founded by treaty between nations, with member states represented by delegates from member governments. The treaty that establishes the creation also establishes the organs, functions, and authority. In most cases, the organization has legal personality and is granted specific advantages under international law. One example is the World Tourism Organization (WTO).

Interline

Anything involving two or more passenger or cargo carriers, especially in air transportation, such as interlining, which is a voyage that involves more than one airline. Normally, such partnerships are managed by agreements between the carriers, which allow one carrier to include the services of another carrier in its tickets while sharing revenue.

Intermediary

For example, a travel agent selling a tour operator's holidays (vacations) to tourists or simply a person or organization acting between parties.

Intermodal

Anything involving two or more modes, e.g., in transport fly-cruise, fly-drive, fly rail represent intermodal journeys. One replacing another is referred to as Intermodal substitution.

Internal Travel/Tourism

Comprises for statistical purposes, from the point of view of a country, domestic travel/tourism plus inbound travel/tourism, e.g., French residents' and foreign visitors' travel/tourism in France.

International Airport

Any airport designated as an airport of entry and departure for international air traffic by the contracting state on whose territory it is located, where customs, immigration, public health, animal, and plant quarantine, and similar procedures are carried out [International Civil Aviation Organization (ICAO)].

International Convention on the Travel Contract

In 1970, an international agreement was signed in Brussels to standardize various legislative regulations in order for tourists to have the most information and protection possible when dealing with travel agencies and tour operators.

International Fare Receipts

For statistical reasons, any payment of sums owing by non-resident tourists to carriers registered in the compiling country, whether or not they are traveling to that country. This category corresponds to 'Transportation, passenger services, credits' in the International Monetary Fund's [World Tourism Organization's (WTO)] standard reporting form.

International Hotel (Telegraph) Code

A letter code that was first used for telegraph communications and is now used to decrease the number of words needed to make a hotel reservation while also assisting with language obstacles. Nowadays, there is minimal trace of its use.

International Load Line

A mark on a ship's hull indicating the maximum depth to which the ship can be sunk due to the cargo's weight.

International Passenger Survey (IPS)

Since 1964, the Office for National Statistics has conducted a continuous sample survey by personal interviews at air, sea, and tunnel ports with passengers entering and leaving the country as the primary source of statistics on incoming and outgoing travel and tourism to and from the United Kingdom (UK).

International Same-Day Visitor

A foreign traveler who does not spend the night in collective or private lodging in the country visited for statistical purposes.

experience, but they also serve as a tool for managing visitor flows and reducing damage.

Inter-Rail Pass

Unlimited travel rail ticket offered to persons who have lived in Europe for at least six months and is good for 12/22 days in up to 28 European and North African countries. Those under the age of 26 and those above the age of 26 pay different prices.

Interstate/Intrastate

Terms used particularly in relation to one or more states of the United States of America, Between/within states. For example, an airline or a bus company providing services between two or more states is an interstate carrier, and one operating within a state, an intrastate carrier.

Intervening Opportunity

A geographical notion that denotes the distance between a prospective visitor's house or tourist base and a major recreational resource, particularly one that is vulnerable to, or has been badly affected by, recreational pressure. They are crucial as a tool for controlling tourist flows: Pressure on a more distant destination could be reduced by building or promoting existing intermediate attractions.

Investment Appraisal

An assessment of the expected costs and income of a potential capital project to determine whether it should be pursued. This is a typical method of evaluating the profitability of private sector projects, such as tourism attractions and facilities. It therefore utilizes one or more of the three main evaluation methodologies – payback period, average rate of return on capital invested, and discounted cash flow.

Investment Incentives

Measures taken by the government to encourage businesses to invest, either in general or in specific industry or places. Financial (e.g., grants, loans, loan subsidies) and quasi-financial (e.g., loan guarantees, exchange rate guarantees) incentives are common in travel, tourism, and hospitality, as are fiscal incentives (e.g., tax exemption or reduction). Financial incentives have the primary effect of lowering the recipient's investment outlay; fiscal incentives have the primary effect of lowering running costs; and quasi-financial incentives have the primary effect of securing the investment.

Investment Intensity

The link between fixed assets like land, buildings, and equipment and current assets like cash, debtors (accounts receivable), and stocks (inventories). The majority of transportation and lodging companies have high investment intensities (i.e., a large proportion of total investment in fixed assets), which adds to high fixed costs of operation due to depreciation and other property-related expenses. On the other hand, tour operations and travel agencies have low investment intensities.

Investors in People (IIP)

Organizations pursuing commercial success through the ongoing growth of their management and personnel are eligible for a British accreditation scheme. It is therefore based on a national standard and provides indicators and a baseline against which an organization's development can be measured.

Invisibles

Receipts and payments from services (as opposed to products), investments overseas, private transfers, and government transactions with foreign nations are included in the current (as opposed to capital) balance of payments statement. Further, international travel/tourism receipts and payments, which include spending in the countries visited, are invisible exports and imports that appear as separate items in the current account. Thus, international fare payments are included in the current account with transportation.

Iron Curtain

A term coined by Winston Churchill in a speech in 1946 to describe the gap (which lasted until 1989) between the Soviet Union and its affiliated Communist republics in Eastern Europe on the one hand, and Western European countries on the other.

Although there was a huge volume of travel and tourism between Eastern and Western Europe, the Iron Curtain served as a barrier between East and West in travel and tourism, as well as other aspects of life, for the most of the period following WWII. Therefore, following the fall of communism in Eastern Europe in the late 1980s, Europe enjoyed a period of rapid East-West tourist expansion.

Iso-

A prefix used for lines drawn on a map connecting points of equal value or quantity, e.g., isobar (atmospheric pressure), isohel (sunshine), isohyet (rainfall), isotherm (temperature).

Isthmus

Connecting two larger land areas, a narrow strip of land, with water on each side (Figure 41).

Figure 41. The isthmus of Corinth.

Source: Image by Flickr.

Itinerary

Showing dates and times, mode of transport, places visited and activities, it is a description of a journey.

J

Jacuzzi, Jacuzzi Bath

A type of bath, or a trademark for a type of bath, with jets which circulate the water and keep it bubbling. The health club has two Jacuzzis and a whirlpool.

Jam

- Prepare a sweet food with fruit and sugar for catering. There are small pots of jam and honey on each table. Individual packs of butter and jam are available at the buffet.
- A stumbling barrier that inhibits something from moving verb to stop operating, to be stumbling block. The traffic lights were out, and the road was clogged for miles.

Jeep

Road travel is a trademark for a rugged four-wheel-drive vehicle designed for harsh terrain travel. The jeep convoy moved slowly across the bridge.

Jeep Safari

An organized sightseeing tour or excursion in a jeep or other four-wheel-drive vehicle.

Jelly

- Food, a semi-solid foodstuff, especially a sweet food consisting of gelatin, water, and fruit flavoring, for example. Fish fingers and chips were served to the kids, followed by jelly and ice cream. (*Note:* Jellies is the plural form.) In the United States, this is commonly referred to as Jell-O).
- Catering, a form of preserve created from fruit juice that has been cooked with sugar and served with mint jelly roast lamb.
- American food, fruit, and sugar are combined to make a delicious preserve. Sandwiches with peanut butter and blackcurrant jelly are her favorite. (*Note:* British English is a shambles).

Jet Lag

A fatigued state after flying across numerous time zones in a jet aircraft After the travel from New York, she had terrible jet lag.

Jewelry

Wearable ornaments made of valuable stones, precious metals, or fake stones. The thief took all of her valuables. (*Note:* In the United States, the word Jewelry is spelled differently).

Job

- A duty, a specific piece of work that must be completed to do very effectively to perform odd chores to perform general work He helps us out around the hotel with odd duties.

Job Evaluation

Examining various occupations within an organization to determine what skills and qualifications are required to do them.

Job Satisfaction

A feeling of being happy in their place of work and pleased with the work they do which employees have.

Job Security

A feeling of having a right to keep their jobs, or that they can stay in their jobs until they retire, which employees have.

Jog

A leisurely run, especially if it's for exercise. Every morning, she goes for a jog.

- To jog, He was going for a jog around the park. Every morning, she goes for a half-hour jog;
- To lightly shake or push My neighbor grabbed my arm and made me spill my drink when the turbulence began.

Jogging Track

Without having to run on the street, it is a SPORT or a track in the grounds of a hotel, health farm, etc., where guests can go jogging (Figure 42).

Figure 42. Po Hong Park in Po Lam, Hong Kong.

Source: Image by Wikimedia Commons.

Join

- To assemble something – the two bedrooms have been connected to form a suite;
- To become a member of an association or a group to join an association or a group. In Cairo, two more persons will join the gang.

Joint

- Meat is number one. A cut of meat, particularly one suitable for roasting a cow joint a bacon joint, a cut from the joint, and a slice of roast flesh a hamburger joint adjective;
- A low-class club or restaurant (informal);
- A business in which two or more entities are connected;
- For a group of two or more persons.

Joint Venture

An extremely large business project in which two or more businesses join together to oversee the project, sometimes by forming a new joint venture.

Journey

A period of move in which you travel from one location to another. For example, he planned his trip to visit all of Europe's capitals in two weeks. We had to walk the rest of the way since our automobile broke down.

Junk Food

Nutrient-poor foods, such as high-fat manufactured snacks, to be consumed in between or instead of meals. They simply sat in front of the television and ate junk food during their vacation.

Jainism

The East Indian Sect's Asian religious system, which has ideas that are quite similar to Buddhism and dates from the sixth and fifth centuries BC.

Japanese Encephalitis

Inflammation of the brain caused by a virus that can be lethal. It is spread by mosquitoes that have bitten sick farm animals and is found primarily in rural regions and during the monsoon season in Southeast Asia. Moreover, travelers staying in high-risk locations for more than a month should get vaccinated, and all tourists should avoid mosquito bites.

Japanese Religions

Japan's religious traditions are founded on Buddhism, Confucianism, and Christianity, as well as a local history of Shinto as emperor-worshipping religion, which became the official state religion in the nineteenth century with a focus on state loyalty as a religious duty. However, the post-World War II constitution encourages religious liberty and the separation of church and state.

Jet Boating

Jet boats (rapid moving boats with a shallow draft designed by a New Zealander), which offer participants an exhilarating experience frequently in attractive locations, are a popular form of recreation and a major tourist/visitor attraction on rivers.

Jet Lag

A syndrome induced by the disturbance of the human body's natural rhythms and experienced by travelers flying great distances across multiple time zones. It is therefore most typically seen in sleeplessness and other illnesses, and it is more common on eastward trips than westbound flights, but the effect differs between people.

Jetty

Usually constructed to protect the shore or harbor or to serve as a landing stage, it is a structure projecting into the sea, lake, or river.

Jeep

For a strongly built four-wheel-drive vehicle, it is a trademark used for traveling over rough ground.

Jeep Safari

It is an organized sightseeing tour or excursion in a jeep or other four-wheel-drive vehicle (Figure 43).

Figure 43. Jeep safari, Africa.

Source: Image by Pixabay.

Jitney

A motor vehicle similar to a bus that transports passengers for a nominal charge over a flexible route, picking up and dropping off passengers anywhere along the way. It is a popular mode of public transportation in places like the Caribbean Islands.

Job Rotation

A method of multi-skilled employment that involves rotating employees through a variety of roles to eliminate boredom, boost motivation, and improve staffing flexibility. In a hotel, for example, the same employees may work as maids, barmaids, and waitresses.

Job Sharing

A system in which a full-time job is split between several people, each working part-time (PT).

Jogging Track

A jogging track in the grounds of a hotel, health farm, or other facility where guests can jog without having to run on the street.

Joint Venture

A contractual agreement involving two or more different parties in the joint ownership of business units and joint participation in their financial outcomes, as well as the parties' involvement in the venture's management and operation, as stipulated in the agreement. This is a popular structure in the hotel and property industries, and it is becoming more common in public-private partnerships, such as resort development.

Judaism

The profession and practice of Jewish religion; the monotheistic belief system detailed in the Bible and the Jewish Law taught by ancient Hebrew priests as a revelation of the divine will.

Jump Seat

If a flight is full, a passenger may be allowed to utilize a stowable seat in an aircraft, usually in the cockpit.

K

Key Card

- An electronic card that serves as a key to a guest's room and is handed to them at check-in;
- A card given to a guest upon registration that shows the room number and may be required for identification purposes.

Key-Drop Box

Guests can drop their keys in a box at reception as they depart the property.

Key Money

A fee paid when you take possession of the keys to a rented apartment or business.

Key Rack

At the front desk of a hotel, a board with hooks or a set of pigeonholes is used to storeroom keys.

Keelage

Shipowners pay a charge to be allowed to dock in specific ports and harbors.

Kibbutz

Working holidays (tours) and sometimes homestay lodging for paying tourists is common in Israeli communal living establishments.

Kidney

- In animals, one of a pair of organs that removes impurities from the blood;
- Dish Grilled kidneys with bacon were used as a food source for this organ.

Note: Lamb's, ox, and pig's kidneys can all be cooked in a red wine sauce or utilized in kebabs. Steak and kidney pies or pudding are the most well-known English meals that use kidneys (Figure 44).

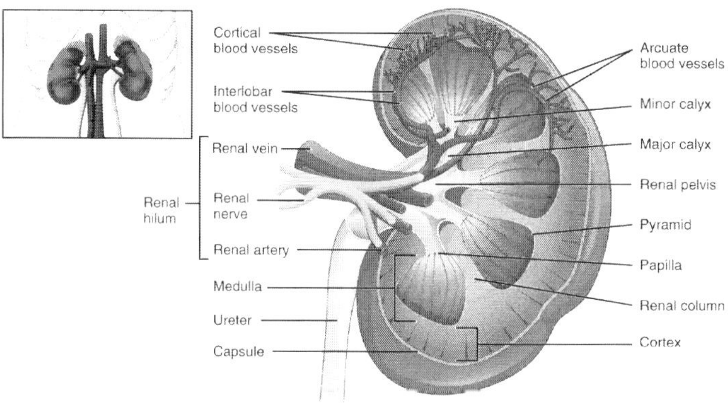

Figure 44. The kidney.

Source: Image by Wikimedia Commons.

Kilometer (Km)

A metric unit of measurement equal to 0.62 miles (3280.84 feet) in length. A kilometer is split into 1,000 meters. In area terms, one square kilometer represents 100 hectares of 10,000 square meters.

King (Size)

A term used to describe a double bed that is extra wide and extra-long, with a minimum width of 72,380 in (1,803,200 cm).

Kitchen

A place where food is cooked before it is served. In the hotel kitchens, the inspector discovered cockroaches.

Comment: In a large restaurant, the kitchen is divided into brigades: the chef (or chef de cuisine) is in charge of all departments of the kitchen; several sectors (each under the supervision of a chef de partie) include sauces, vegetables, pastries, and so on.

Orders are taken from the waiters and shouted out to the various sections by a clerk (the 'aboyeur'). Assistants (or commis chefs), porters, plate washers (plongeurs), and other kitchen personnel will be present.

Kitchen Clerk

The worker in the restaurant who announces the waiter's order to the cooks and attaches the waiter's written order to a table-specific hook. Also known as Aboyeur.

Kite

Players in kitesurfing or kiteboarding use a huge, crescent-shaped instrument made of fabric stretched over a light frame to provide propulsion and lift.

Kiteboarding

A sport in which participants use a kite tied to their bodies to provide power and lift while riding skateboards or snowboards.

Kiwi

A flightless bird that is generally known as New Zealand's emblem; a slang term for a New Zealander.

Kleenex

A logo for a handkerchief made of paper. In the restroom, there is a box of Kleenex. (*Note:* There is a plural form Kleenexes for several handkerchiefs.

Kloofing

The adventurous sport of climbing, swimming, and leaping to follow the course of a stream through a canyon.

Knife

A tool with a sharp blade that can be used for cutting and spreading.

Comment: All chefs agree that their knives are the most significant component of their equipment (each owns a set of multiple knives that they carry from job to job). Knives that are forged are the best (i.e., the blade, bolster, and central portion of the handle are all composed of the same metal).

A good knife must have a strong handle to balance easily: the tang is the central metal component of the handle to which hardwood or plastic grips are riveted. The tang is the component of the handle that provides weight (Figure 45).

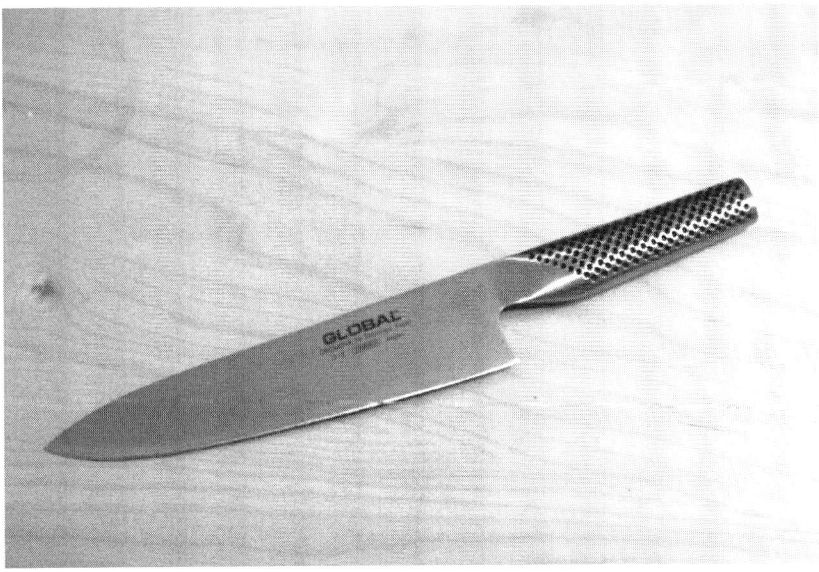

Figure 45. The kitchen knife.

Source: Image by Wikimedia Commons.

Knot

One international nautical mile (1.852 kilometers or 6,076 feet) per hour is a unit of measurement.

Knowledge

The common facts or information that individuals are aware of Knowledge of the London Street plan, which taxi drivers must memories and on which they are tested prior to receiving their license (informal).

Koori Tourism

Tourist activity that involves a cultural feature of the indigenous people of south-east Australia known as the Kooris. Unlike northern Australian aborigines, who live a predominantly traditional lifestyle. Kooris are more prevalent in metropolitan areas, and Koori tourism frequently comprises purpose-built 'culture centers' that depict history and contemporary Koori life.

Korma

Finance is a pleasantly spiced dish of meat, fish, or veggies cooked in a cream or yogurt sauce in South Asian cuisine.

Kyle

A small canal between two islands or between an island and the mainland, as seen, for example, in Kyles of Bute, a popular tourist destination in Scotland.

Kyoto Agreement

Further to the Earth Summit, developed countries agreed to legally enforceable targets to cut emissions of the basket of six major greenhouse gases at the Kyoto Conference in Japan in 1997.

L

Labeling of Alcoholic Beverages

The April 1987 European Commission Directive addresses the statement of alcoholic intensity by percentage in the labeling of alcoholic beverages for sale to the final consumer in member countries.

Labor Day

On the first Monday in September, the United States celebrates its national holiday as a public holiday.

Labor Force Survey (LFS)

The British household sample survey was done every three years until 1983, then annually until 1991, with interviews taking place in late spring and a sample size of approximately 60,000 homes. The Survey has been conducted quarterly since spring 1992. Hence, estimates of employees, self-employed people, and the jobless at a national and regional level are provided.

Labor-Intensive

When compared to other productive resources, such as capital, an economic activity is labor-intensive when it necessitates a significant labor contribution. While tourism is often thought to be labor-intensive, tourist-related businesses such as transportation are capital-intensive. However, a given level of tourist expenditure generates more jobs than a comparable amount of consumer spending in many other economic sectors.

Lagoon

A low sandbank or coral reef separates a shallow stretch of coastal water from the open sea. (b) A body of water surrounded by an atoll (Figure 46).

Figure 46. The coral lagoon near Armier Bay, Mellieha.

Source: Image by Wikimedia Commons.

Lanai

A room with a patio or balcony, preferably with a picturesque view or overlooking a planted area, is referred to as a veranda in Hawaiian.

Land Resources Depletion

As a result of human activity, minerals, fossil fuels, rich soil, forests, wetlands, and wildlife are depleting at an accelerating rate. Therefore, potential loss of new destinations, deterioration of current ones, and increasing fuel prices leading to operating price rises in tourism-related businesses are all serious long-term repercussions for travel, tourism, and hospitality.

Land Use Planning

The process of zoning land for certain uses based on environmental, social, and economic factors in order to suit current and future needs, such as industrial, housing, or recreation. When addressing visitors' demands is crucial, such as in resort regions, tourism may be subsumed with recreation or represented as a separate aspect in such planning.

Land Use Planning Systems (UK)

Plans developed by county planning authorities, a few unitary and National Park authorities, as well as district councils in England and Wales established out broad policies; local development projects planned by some unitary and National Park authorities, and district councils include comprehensive guidance for growth expected to begin within the next 10 years. Moreover, in overall compliance with structure plans; unitary development projects incorporate broad policies and detailed guidelines for metropolitan districts in England and Wales. Hence, in Scotland, regional, and islands administrations develop structure plans, whereas districts, general planning, and islands agencies prepare local plans.

Landau

A four-wheeled horse-drawn carriage with a two-part top that may be entirely closed, half- or fully opened, and is sometimes used for town and city touring, with German origins.

Land-Bridge

Travel by sea from/to continental Europe via Great Britain (GB) is referred to as "sea travel" in Ireland.

Landlocked Country

Switzerland in Europe, Zambia in Africa, Nepal in Asia, and Paraguay in South America are examples of countries without a coast and hence direct access to the sea.

Landing Card

Passengers are given a card to fill out and hand over to immigration and passport control.

Landscape

A term used to express the completeness of the scenery or the territory defined by a certain sort of scenery, derived from a Dutch term for the representation of scenery in a painting. The landscape garden movement in England in the 18th century, primarily concerned with the names Humphrey Repton and Capability Brown, sought to create a harmonious whole in the park surrounding a vast country house; an essential feature of community gardens and affiliated parks, such as Blenheim Palace, which are now important tourist/visitor attractions.

Lapland

Northern Finland, Norway, Sweden, and Russia's Kola Peninsula, mostly in the Arctic Circle, are home to the Lapp people, who speak their own language.

Late Arrival

Term used by hotels to describe a guest that comes or is anticipated to arrive after the hotel's latest check-in time, usually after the hotel receives assured payment and/or prior notification of late arrival.

Late Cancellation

Cancellation of a reservation for lodging, transportation, or another service after the time restriction has passed. If the canceled room or seat cannot be resold, there may be a fee.

Late Check-Out

Checking out of a hotel after the check-out time, which is usually but not always noon. Although hotels may charge a fee, it is normal practice to extend a guest's check-out time if the room is not needed for another guest.

Latin America

South America (excluding Guyana, French Guyana, and Suriname), Central America (except Belize), Cuba, Dominica, and the Spanish-speaking islands of the West Indies (WI) make up the New World's Spanish-speaking countries. The most major international tourist destinations in the region are Mexico and the Caribbean, which owe their tourism boom to their closeness to the vast and wealthy North American market.

Latitude

The angular distance north or south of the equator is expressed in degrees, minutes, and seconds from the center of earth. Each degree is divided into 60 minutes and each minute into 60 seconds, with 90° between the equator and each pole.

Law of Diminishing Returns

This economic law asserts that adding more units of an input while keeping other inputs constant will lead to less than a proportionate rise in output sooner or later.

Layover Scheduled

Interruption of a voyage, usually overnight, at the passenger's request or because of a lack of connecting services, for example. Also known as a layover. Crews, particularly airline crews on long-haul flights, are subject to the same terms.

League

A distance measurement that varies by country but is usually three miles, with the land league in the UK and the US being three statute miles and the nautical league being three nautical miles.

Leveraged Buyout

The purchase of a firm by another company with borrowed capital, in which the acquiring company pledges its own assets as collateral for a loan that will be returned from the acquired company's cash flow.

License

Generally, a government agency or another regulatory entity grants authority to accomplish what is defined by the license, which would be prohibited without it. Various modes of transportation, as well as other travel and tourism activities, require a license in most countries.

Life Cycle

The progression of life's stages from childhood to old age. As a process in the family history, which includes, for example, the influence of children on family behavior, it is frequently applied more meaningfully to families than to individuals. The notion can be used to categorize consumer markets in the travel, tourist, and hospitality industries. Moreover, the life cycle idea, as applied to products and, in the case of travel and tourism, to destinations, should be distinguished, as they are viewed to pass through stages of exploration, development, consolidation, stagnation, and either decline or renewal.

Lifestyle

A term used to define an individual's or a group's way of life as expressed in consumption and other behavior, which is important in market segmentation.

Line Route System

A system of airline routes in which planes take off from a base airport and stopover at a number of intermediate points to refuel or pick up travelers before continuing on to their final destination.

Liner

A large seagoing ship that operates on a regular schedule for passengers and/or cargo on specific routes.

Linkages

The amount to which industries, rather than importing products and services, employ products and services accessible from other sectors of the same economy. Tourism is said to have strong links with other sectors of the economy if the majority of goods and services consumed by the industry are produced domestically. Moreover, regions and other geographical areas are subject to the same idea. In general, the smaller the economy, the fewer the firm-to-firm links and the higher the risk of leakage.

Livery

The expression "distinctive visual corporate identity" refers to an airline's distinct visual corporate identity that it displays through design and house style on its aircraft, uniforms, merchandise, printed content, and other locations where customers may view it. The need to distinguish an airline from other airlines involves the upkeep and advancement of its corporate identity, and livery is used as a non-price competition tool (Figure 48).

Figure 48. A Himalaya airlines airbus A320-214 at Tribhuwan International Airport.

Source: Image by Wikimedia Commons.

Load Factor

The ratio of transport capacity sold to capacity used as a proportion of total capacity. The passenger load factor (also known as passenger seat occupancy) is the ratio of passenger kilometers to seat kilometers, while the revenue load factor is the ratio of load ton-kilometers to capacity ton-kilometers. In hotels and other lodging enterprises, the load factor is a measure of utilization similar to occupancy.

Load-Ton Kilometers

The product of the revenue load in tones and the distance in kilometers over which it is transported is used to determine transport sales.

Location Factor

P. Sargant Florence is credited with inventing a measure of industry concentration in a region that compares the proportion of the economically active population in a specific area engaged in a given industry to the proportion engaged in the country as a whole. As a result, location considerations provide a metric for comparing different areas.

Locator Map

A map that appears in guide books and other travel reference books, usually covering a town, city, or other local area and highlighting attractions and lodging. Instead of the degrees of latitude and longitude found on traditional maps, grids are employed to aid localization.

Lock

A section of canal or river surrounded by gates into which vessels can enter to be lifted or lowered by water being let in or out. Boats can therefore, move from one canal or river reach to another at different levels in this manner.

Lodging Industry

The word "hotel industry" is commonly used in the United States to refer to the hotel industry, but it is also used in a broader sense to refer to all or most other commercial hospitality establishments, including guest homes and condominium.

Logo

A distinctive design, symbol, or other representation of a company's or organization's name is used to identify and distinguish it from competitors.

Lomé Convention

The European Economic Community (EEC) and developing countries of Africa, the Caribbean, and the Pacific (ACP States) signed an agreement in 1975 in Lomé, Togo, that offered duty-free access for most ACP countries' exports to the EEC as well as financial and technical assistance.The most recent Convention, Lomé IV, covered the period 1991–2000. The EU and ACP presidents of state and governments signed the Cotonou Agreement, a new 20-year partnership agreement with 77 ACP countries, in June 2000 in Cotonou, Benin.

Long-Distance and Short-Distance Travel/Tourism

Differentiation is especially important in civil aviation because of differences in aircraft types, operating, and traffic managing tactics, and marketing. Travel/tourism between/within continents or between/within global areas is sometimes used interchangeably. However, the distinction is best made based on the measurable distance traveled. The consumers' association's holiday which, for example. Long-haul travel is defined as flights lasting more than five hours.

Long Vacations (Vacations)

Holidays (vacations) of four nights/five days or longer outdoors are defined as holidays (vacations) in most national holiday (vacation)/travel/tourism surveys in Europe.

Long Service Leave

For Australian employees who have worked for a single public sector employer for 10 years, they are entitled to 13 weeks of paid leave. This benefit has allowed Australians to take long vacations (vacations) away from home, allowing them to explore Australia and/or go too far-distance places such as the United Kingdom (UK) and the rest of Europe.

Long-Distance Footpath

A long, straight pedestrian path that usually runs across open or forested hill area. Along the route, lodging can be found in adjacent communities, huts, or tents. The National Parks and Access to the Countryside Act 1949 was passed in England and Wales, while the Countryside (Scotland) Act 1967 was passed in Scotland. In North America, it is known as a trail, such as the Appalachian Trail, which stretches over 2,000 kilometers (1,200 miles). However, it is unknown what percentage of people walk the entire length of a trail or footpath.

Longitude

Any place on the earth's surface expressed in degrees, minutes, and seconds east or west of the zero meridian, which passes through Greenwich, England.

Loss Leader

A good or service sold at a low price to entice clients to buy more items, a strategy popular in supermarket retailing but also employed in restaurants and other eating facilities, for example.

Luncheon Vouchers

Employees are given tickets to use as payment for food in restaurants that accept the coupons. Specialty companies, such as Luncheon Vouchers Ltd in the UK, give vouchers to employers that can be redeemed for cash at restaurants.

Lunch Hour

During the lunch hour, a series of lunchtime concerts are held in the public gardens. Throughout lunchtime or during the lunch hour, the office is closed.

Luxurious

Very relaxing. Luxurious carpets and fixtures have been installed in the guest rooms. Business-class does not offer the same level of comfort as first class.

Luxury

It's a wonderful thing to have a black market in luxury items because it's an expensive thing that isn't necessary.

Luxury Suite

Hotels are a collection of luxurious rooms in a hotel, apartment building, or ship.

Lyophilize

Freeze-drying is a method of preserving food by fast freezing it and then drying it in a vacuum.

M

Macaroni

Short, thick spaghetti tubes, frequently served with a cheese sauce (Figure 49).

Figure 49. Macaroni pasta.

Source: Image by Wikimedia Commons.

Macchiato

A cup of espresso coffee with a dash of steaming milk on top.

Mackerel

A dark-fleshed sea fish that can be grilled or smoked, as well as canned. As a beginning, we had smoked mackerel.

Macrobiotic

A vegan diet consisting of grains, seeds, and organically cultivated fruit and vegetables is supposed to extend life and balance physiological systems.

Maggot

The larva of an insect, such as a bluebottle, warble fly, or fruit fly, with a soft body and no legs. Maggots had infested the meat.

Maid

- A girl or woman who helps around the house, particularly with cleaning and meal serving. The towels had not been changed by the maid;
- Hotels are similar to chambermaids. Cleaning is done every day in the chalet by a maid.

Maiden Flight

An aircraft's first flight or a ship's first voyage on her maiden journey across the Atlantic in 1912, the 'Titanic' sank.

Mail

A system for sending letters and parcels from one location to another in order to mail a letter. The check was misplaced in the mail. The invoice was mailed out the day before yesterday. Therefore, mail can take up to six weeks to reach some Pacific islands. Also termed post by mail using the postal services, not sending things by hand or by sender by sea mail sent by post abroad, using a ship we sent out the order by first-class mail by the costliest mail service, intended to be faster.

Letters that have been sent or received Is your mail still on its way? Your check arrived in the mail yesterday. As soon as my letter arrives, my helper opens it. The receipt arrived in the mail this morning.

To mail a letter confirming the reservation, the verb is to send anything via post. Last Wednesday, we sent out our confirmation.

Mailbox

- One of several boxes in a large building where incoming mail is received, or a box where letters and packages to be posted are placed;
- An e-mail address for receiving messages.

Maastricht Treaty

The Treaty on European Union (EU) was signed on February 7, 1992, in Maastricht, Netherlands, by delegates of the 12 countries of the European Community (EC), establishing EU and introducing EU citizenship for people of member countries. Therefore, the treaty intends to improve intergovernmental collaboration in economic and monetary concerns, establish a shared foreign and security strategy, and promote justice and home affairs cooperation.

Macadam

Road-building term named for Scottish road engineer John Loudon McAdam, one of the three major road-makers of the late eighteenth century (the others being John Metcalf and Thomas Telford). Moreover, the phrase refers to the compacted shattered stone that forms a hard surface, and the roads that are constructed in this manner are known as macadam roads or macadamized roads. Hence,armacadam is a later development that refers to a substance made of stones or iron slag covered with tar or tar and creosote and used, for example, for airport runways.

Maglev

Magnetic levitation is a technology for suspending a train above the track. It is a high-speed mode of surface transportation that is used in the United Kingdom (UK), for example, connecting Birmingham International railway station and Birmingham Airport and at London Gatwick Airport.

Maiden

Term used in travel and tourism (and elsewhere) to describe something that is the first of its kind or used for the first time, especially in transportation. Thus, maiden voyage refers to the first trip of a new ship, maiden flight refers to the first flight of a new aircraft type, and maiden route refers to the first use of a new route, however the latter is more usually referred to as inaugural flight.

Main Activity Index

T.C. Huan developed a scale to assess the relative importance of diverse tourist activities [Huan, T.C., and O'Leary, J. T. (1999) Measuring Tourism Performance, Champaign, IL: Sagamore Publishing].

Main Holiday (Vacation)

The definition "main holiday" is used in some holiday (vacation) survey data, such as the British National Travel Survey (BNTS): "Where only one is taken, this is the important holiday; if two or more are taken, the main holiday is the longest or, if two or more are of similar duration, the one in or closest to the peak summer period."

Mainstream Tourism

Tourism to well-known vacation areas using common transportation and without the need for additional arrangements.

Maintain

- To maintain something in good working order in order to maintain good relations with customers in order to keep in touch with a courier overseas via phone;
- To maintain the status quo despite the recession, the restaurant has kept the same level of business.

Maintenance

- Efforts to keep things running smoothly, such as contact maintenance (e.g., supply replenishment).
- The task of maintaining the proper operation of a machine.

Maize

A grain crop that is commonly farmed. Corn is another name for it (*Note:* The US English is corn).

Comment: In Europe, just a small percentage of the harvest is sold as 'corn on the cob' for human consumption.

Maize is the only grain crop that was brought from the New World to the Old World, and its moniker, Indian maize, comes from the fact that it was grown by American Indians before European settlers arrived. It is therefore the most widely grown crop in the United States, and it is the most often consumed food in Mexico, where it is roughly ground into flour and used to make tortillas (Figure 50).

Figure 50. Maize.

Source: Image by Wikimedia Commons.

Major Operated Departments

Rooms, food, and beverages are the primary revenue-generating hotel operations, as opposed to smaller run departments, rentals, and concessions, which are bundled for accounting and control purposes. [Hotel Uniform System of Accounts].

Malaria

Cerebral malaria is one of the most prevalent travel-related disorders, causing high fever, jaundice, and coma. Infected mosquitoes, which bite after dark, spread the disease. The Mediterranean, Southeast Asia, Sub-Saharan Africa, and South America are the main sites of infection. Moreover, antimalarial medications should be used before, during, and after travel, as well as avoiding mosquito bites.

Malt

- A grain that has been sprouted and then dried in preparation for making beer or whisky;
- In the United States, it's the same as malted milk.

Malted Milk

- Malt-flavored dried milk powder;
- A drink made by combining this powder with fresh milk.

Management

'Management, as a subject, may be viewed as the process of employing material and human resources to achieve stated objectives, involving the activities of planning, organizing, directing, coordinating, and controlling,' according to the British Standard. Therefore, the functions stated above are performed by management as a group of persons. Top, medium, and lower management are the most common levels of management authority [BS 3138 (1979)].

Management Accounts

Financial information for a management, such as sales, expenditures, credit, and profitability, provided for him or her to make judgments.

Management Committee

A committee in charge of a club, a block of apartments, or a pension fund, for example.

Management Consultant

A business advisor is someone who provides guidance on how to run a business.

Management Contract

A contract between two parties for the provision of organizational and operational competence, in which one party (the owner or principal) hires another (the contractor or agent) to run a firm for a fee. It further specifies the parties' respective responsibilities, rights, and obligations, as well as the payment of expenses, management fees, and/or profit-sharing. Various tourism institutions, particularly hotels and catering establishments, use management contracts.

Management Training

Training for employees who desire to be managers that requires them to research problems and devise solutions.

Manager

- A firm's head of department, such as a personnel manager, a sales manager, or a purchasing manager;
- A manager of a hotel, inn, branch, or store. Mr. Smith is the proprietor of our neighborhood tavern. Our New York hotel's general manager is in London for a series of meetings.

Mandolin

A kitchen equipment with adjustable blades in a frame for slicing vegetables.

Mango

A huge, yellow, or yellowish-green tropical fruit with a delicate orange pulp encircling a very large flat seed (Figure 51).

Figure 51. Image showing mango.

Source: Image by Flickr.

Manila Declaration

In 1980, the World Tourism Conference in Manila, Philippines, adopted a declaration. It clarifies the nature of tourism and its function in a dynamic and changing world, as well as the states' responsibilities for tourist development and enhancement. Moreover, other agreements and announcements on problems relating to tourism development are based on the Declaration.

Manpower

Personnel shortage, manpower shortage, lack of workers are all terms used to describe a scarcity of workers in a business or industry.

Manufacturer's Recommended Price (MRP)

The suggested retail price for a product set by the manufacturer, though this is frequently decreased by the store. Also known as the suggested retail price.

Marginal Pricing

Method of determining a product's selling price based on variable production costs plus a margin, while excluding fixed expenses.

Marina

A small sea, lake, or river port with motor and sailing boat docking services, as well as maintenance and supply services. Most marinas offer both long- and short-term berths, as well as boat charter.

Marinade

A wine-and-herb combination that is steeped in before cooking meat or fish, or soak meat or fish in wine and herbs for catering Before cooking, marinate the meat for 12 hours.

Marinated

Before cooking, the catering was soaked in a mixture of wine and herbs. Anchovies are marinated in-house rather than being cured.

Marine Park

A park built on the seabed where visitors can see marine life from underwater observation rooms.

Market

In economics, a market is a collection of transactions between the vendors and buyers of a product; a market is characterized by the commodity, the sellers who supply it, and the buyers who exert demand for it. As a result, tourist markets are characterized in terms of tourist items, suppliers, and tourists who purchase them. Moreover, buyers are considered as constituting the market in business, either the actual (existing) market, which includes people who buy the product now, or the potential market, which includes those who may buy it in the future. As a result, tourist markets concentrate on current and potential visitors.

Marketing

Packaging and advertising are examples of techniques used to sell a product.

Marketing Department

A department of a firm that focuses on different ways to promote a product.

Marketing Mix

Marketing is a term that refers to a collection of marketing factors, such as price, distribution, and advertising.

Market Share

The proportion of a whole market that a company's sales cover.

Mark-Up

- An increase in a product's price in June, we implemented a 10% pricing increase across the board;
- A quantity added to the cost price to give the selling price we function to a 3.5 times mark-up or to a 350% mark-up we take the unit price and multiply by 3.5 to yield the selling price.

Marmalade

Orange jam or other citrus fruit jam (lemon, grapefruit, etc.), Marmalade prepared with bitter oranges (Seville orange marmalade).

Comment: Marmalade is only consumed on toast for breakfast.

Marquee

A tent used for social gatherings or exhibitions in most nations; in the United States, primarily a long narrow tent with no sides used for sheltering walkways.

Marquise

- A chilled French dessert made of whipped cream and fruit-flavored ice;
- A rich chocolate mousse or a spongy chocolate cake, or a blend of the following, is the base of this French dessert (Figure 52).

Figure 52. Chocolate marquise.

Source: Image by Flickr.

Marrow

A melon-like plant with a huge green cylindrical vegetable with spongy white flesh. (Marrows are typically taken when they are quite little, and are referred to as courgettes or zucchini at this point).

Marshall Plan

General G.C. Marshall, then US Secretary of State, was the inspiration for the European recovery program (ERP), which saw the US send help to European countries between 1948 and 1952. Therefore, the Organization for European Economic Cooperation (OEEC), which later became the Organization for Economic Co-operation and Development, was in charge of the initiative (OECD). Hence, recognizing the importance of tourism to their economies, several countries, including Austria and Italy, invested a significant percentage of the ERP monies in their hotel businesses.

Marzipan

A paste made of ground almonds, sugar, and egg that can be used to coat a fruit cake before icing or to produce individual desserts.

Masala

A paste made from a variety of spices used to flavor South Asian foods, or a dish flavored with this substance.

Mashed Potatoes

Peeled, boiling potatoes that have been crushed with butter and milk till a soft cream has formed. Creamed potatoes are another name for them.

Maslow's Needs Theory

Maslow's triangle of the human hierarchy of needs can be linked in terms of motivating factors to the need for traveling, tourism, and hospitality, according to a theory of individual growth and motivation proposed by behavioral scientist Abraham Maslow in 1954.

Mass

- A large group of individuals;
- A large quantity or number of things. We have a large number of letters to write, or a large number of letters to write. Following the television advertising, they received a flood of requests.

Massage

Rubbing the body in order to reduce pain or enhance circulation. She had a massage and a soak in the sauna. To rub someone's body to ease pain or enhance circulation is the verb rub.

Mass Media

Entertainment is a term that refers to a group of different ways of transmitting information to the public, including television, radio, and newspapers.

Mass Tourism

Term used to define large-scale tourism engagement in industrialized countries in the second half of the twentieth century, as opposed to earlier times and the situation in developing countries, as well as restricted participation in activities such as mountain climbing and sailing.

Master Franchise

The primary franchise for a region, from which sub-franchisees are issued.

Master of Ceremonies (MC)

Entertainment refers to a person who introduces the speakers at a meal or a formal occasion.

Master of Culinary Arts (MCA)

Catering is a four-year award given by the Academy of Culinary Arts to senior restaurant staff.

Maximum Permitted Mileage (MPM)

Maximum mileage between any two points (i.e., airports) mentioned in the tariff manuals, which is approximately 20% greater than the actual direct mileage and gives passengers paying full rates a choice of routings and carriers. For example, the distance between London and Milan is 584 miles, while the MPM is 700. As a result, 1,400 miles per round trip is allowed (although unused mileage on the outward sector cannot be used on the return sector). At the same roundtrip ticket as London–Milan, the customer can fly (and stopover) through Brussels and Geneva or Paris and Zurich.

Mayday

Use an international signal to indicate that you are in trouble. The ship issued a Mayday signal. The captain radioed air traffic control to report the plane was in danger, and the captain issued a Mayday.

Mayonnaise

A cold-dish sauce composed of oil, eggs, and lemon juice or vinegar. Do you want some mayonnaise on top of your salad? We purchased tuna sandwiches with mayonnaise.

Comment: Mayonnaise is consumed with cold boiled eggs (oeufs mayonnaise), cold lobster (lobster mayonnaise), and other cold foods; it's also a base for other dipping sauces like tartare sauce and Thousand Island dressing (Figure 53).

Figure 53. Mayonnaise.

Source: Image by Wikimedia Commons

Meal

- Food consumed at a specific time of day Breakfast, lunch, and supper are included in full board. For a modest fee, you can have your meals delivered to your room;
- Flour that has been coarsely ground.

Measure

- A method of determining the size or quantity of something;
- A serving of alcoholic beverage or wine served by the glass in order to serve smaller amounts of alcohol than permitted by law.

Inspector of weights and measures verb to determine the size or amount of anything to determine the size of a package to be measured for a suit to have your measurements taken by a tailor when designing a suit for you, such as arm length, chest, and waist.

Meat

The meat course is the main dish in a meal, consisting of meat and vegetables, that comes from an animal's body.

Comment: Different varieties of meat have different names than the animals from which they are derived.

Calves give 'veal,' full-grown cows or bulls give 'beef,' pigs give 'pork,' or if salted, 'bacon,' and 'ham,' sheep and lambs give 'lamb,' and deer give 'venison.' Only chickens, ducks, geese, and turkeys produce meat with the same name.

Meat is frequently referred to on menus by its French name: agneau (lamb), boeuf (beef), porc (pork), jambon (ham), and so on.

Mecca

Mohammed's birthplace and a Muslim pilgrimage site in Saudi Arabia. (b) Any location that one wishes to visit in their lifetime. (c) Honeypot's colloquial synonym.

Media

A channel or vehicle of communication, plural form of medium, a phrase of particular importance in the context of advertising. Therefore, maybe expressly developed for advertising purposes (e.g., a poster) or used for advertising in general (e.g., a billboard, the press, radio, and television). National newspapers, radio, and television are examples of mass media channels that reach enormous audiences.

Medical Aid

- Medical treatment for someone who is sick or injured (e.g., first-aid treatment);
- Medical supplies and experts are sent to a disaster-stricken country.

Medical Inspection

An inspection of a workplace to ensure that the working conditions do not make employees sick.

Meet

- To join forces with someone to meet a tour group at the airport and an agent at her hotel. At 2 p.m., we'll meet our relatives at the main station;
- To satisfy the demand for a new service to provide a new service that has been requested we will try to meet your cost we will try to offer a price that is reasonable to you (*Note:* meeting – met). In order to fulfill the demand for a new service to provide a new service that has been requested we will try to meet your price we will try to offer a price that is acceptable to you.

Melba Toast

A cracker-like toast created by cooking a slice of bread once, then slicing it in half and grilling it again fast (Figure 54).

Figure 54. Melba toast.

Source: Image by Wikimedia Commons.

Melon

A sweet, round, or cylindrical fruit with green, orange, or white pulp. As an appetizer, we had melon and Parma ham.

Comment: Melon can be offered as a first course or as a dessert (often with Parma ham). With the seeds removed, the fruit can be cut into wedges or halves. Port can be poured into the hole in a split melon.

Member

- A person who is a member of a group or society;
- Business a company that is a member of a group or society, such as the European Union's (EU) member countries or ABTA's member companies.

Memorandum of Understanding (MOU)

Two or more parties sign a paper stating their desire to collaborate on a common aim. The MOU, which is usually broad in scope and not legally binding (thus the name "understanding"), may be followed by a more formal document later.

Meningitis

Inflammation of the membranes that surround the brain and spinal cord that affects children in particular and can be fatal. Therefore, the main places affected are hot, dry sections of Africa and Asia, but it is not limited to these regions; cases have also been reported in Europe and other parts of the world. Hence, all persons performing Hajj, the annual Muslim trip to Mecca, must be immunized in Saudi Arabia.

Mental Map

An image of an area carried in the head that differs from a real map in that distances and shapes are likely to be distorted, and inaccuracies and omissions may exist. It is therefore essential in a recreational and tourist environment since it influences destination selections by controlling images of where places are and what they're like. Further, in the absence of a guide book or map, it can also help you find your way.

Menu

A printed menu of a restaurant's meal options. Each table has a copy of the morning menu. Every week, the menu varies. Some unique meals are listed on a dedicated board rather than on the menu. The dinner menu starts at £30 a person and goes up from there.

Comment: A three-course meal (lunch or dinner) will typically begin with a hors d'oeuvre, starter, or soup, followed by a main course of fish, meat, or a vegetarian dish, and finally a dessert or cheese. More elaborate menus will include a fish or pasta course in addition to the main course, as well as a dessert and cheese.

It is worth noting that the dessert course appears before the cheese on English menus, whereas the cheese appears before the dessert course on French menus.

Merchandizing

Activities like as display, packaging, sampling, and special offers are used to influence potential buyers once products reach the point of sale (POS), also known as point-of-sale promotion. The fact that most of travel, tourism, and hospitality is consumed on the premises of the vendors creates unique chances for merchandising, such as in hotels.

Merchant

- A wine merchant is a businessman who buys and sells goods in large quantities, especially imported goods, for retail sale;
- A business or store that only accepts a specific type of credit card for purchases.

Meringue

A beaten egg white and caster sugar concoction that was slowly dried in the oven.

Comment: Small individual meringues with cream or fruit salad can be served; the meringue mixture can also be placed on top of a fruit pie before baking (Figure 55).

Figure 55. Homemade meringues.

Source: Image by Wikimedia Commons.

Meridian

On a map, a line connecting the North and South Poles and intersecting the equator at right angles. Meridians are numbered east and west from the zero meridian in Greenwich, England, to 180° in each direction, with 180° East and 180° West overlapping.

Messing

Food, lodging, and recreational activities for servicemen that are all arranged.

Metal Detector

A gadget that raises an alert if it detects metal and is used, for example, in airports, to check if people are carrying guns or other dangerous objects.

Metro

An overground tram system or an underground railway system the metro in Paris.

Comment: The 'metropolitan,' shortened to 'le métro' or 'the metro' in English, was the name given to the Paris underground railway, and it is still used in several other nations, notably Newcastle in the United Kingdom (UK) and Brussels in Belgium.

Underground railways in Germany are known as 'U-Bahn.' The 'underground,' or simply the 'tube,' is the name given to London's underground railway. A 'subway' is the name given to a subterranean railway in North America.

Metropolitan France

French overseas territories, like Martinique and Guadeloupe in the Caribbean and French Polynesia and New Caledonia in the Pacific, are not included in the home nation (which includes Corsica).

Meunière

Fish dredged in flour, fried in butter, then topped with lemon juice and minced parsley, for example.

Meze

In Greece and Southwest Asia, a variety of snacks, such as filled vine leaves, miniature pastries, or grilled sausages, are offered with drinks as an appetizer or light meal.

Mice

Meetings, Incentive Travel, Conventions, and Exhibitions is an acronym often used by the travel industry to refer to group business travel.

Microstates

Small sovereign states, some having inhabitants of less than 10,000 people, such as Caribbean Island states (e.g., Antigua, and Barbuda, Dominica, St Kitts, and Nevis) and Pacific Island states (e.g., Antigua, and Barbuda, Dominica, St Kitts, and Nevis) (e.g., Kiribati, Nauru, Tuvalu). A microstate is defined as a territory with a population of less than 1 million people; this definition encompasses almost 100 island and non-island territories.

Microwave

An oven that heats by the use of ultra-short-wave radiation. To heat the pie, place it in the microwave for two minutes. To microwave a dish is a verb that means to heat a dish in the microwave (informal). Is it possible to microwave the pie a little longer?

Middle East (ME)

The Middle East (ME), strictly speaking, includes countries bordering the Persian Gulf, as opposed to those bordering the Mediterranean, which are known as the Levant or Near East. However, the word is now often used to refer to the countries of southwest Asia and northeast Africa as a whole, encompassing Iran, Iraq, and Arabia, as well as those along the Mediterranean's eastern coastlines, including Egypt. Moreover, most of the countries have great tourism potential, but conflict and terrorist activity have hampered their development, while oil-based income has resulted in an increase in worldwide travel from the region.

Middle Management

A company's department managers are in charge of carrying out the directors' policies and organizing the work of a group of employees.

Midnight Sun

Between mid-May and end-July, the sun or light will be visible over the horizon at midnight in Arctic (higher than 63°30' north) latitudes and Antarctic (higher than 63°30' south) latitudes.

Midweek

It's taking place in the middle of the week. The fares are greater if you fly during the week rather than on a Saturday night. Therefore, a particular holiday package for the middle of the week is called a midweek discount getaway.

Migration

Movement of people from one place (country, region) to another to take up permanent or long-term residency, which differs from tourism, which is a transitory short-term movement of people beyond their typical place of residence and job. As a result, the terms migrant, emigrant, and immigrant have been coined.

Mild

Beverages used to be a variety of brown beer that was less alcoholic and sweeter than the 'bitter' description.

- yogurt-based sauces that aren't as sharp or as strong in flavor are milder. We're going to go with the mildest curry on the menu. Hot on the opposite side;
- In terms of weather, it is not severe. Winters in the south are often milder than those in the north (*Note:* gentler = gentlest).

Milk

A white liquid produced by female mammals for the purpose of feeding their young, particularly milk produced by cows.

Milk Products

Milk and various products made from it, such as liquid milk, butter, cheese, cream, ice cream, condensed milk, and milk powder, that are sold for human consumption.

Millet

A popular grain crop grown in many of Africa's and Asia's hot, dry climates, where it is a staple diet (Figure 56).

Figure 56. Millet grains.

Source: Image by Wikimedia Commons.

Mince

Meat that has been ground into very small pieces or a paste, usually beef. Catering to grind meat into very tiny chunks or into a paste a pound of beef mince or finely diced pork (*Note:* Ground beef is the American English equivalent of minced beef).

Mincer, Mincing Machine

A machine that grinds meat into extremely fine pieces. Contamination can be spread by poorly cleaned mincers. (*Note:* The English spoken in the United States is a grinder).

Mineral Water

Water that originates from the ground naturally and is sold in bottles.

Comment: There are thousands of distinct kinds of mineral water, each with its own set of characteristics. Evian, Vichy, Perrier, Badoit, and others are popular in France; San Pellegrino is popular in Italy; and Spa water is popular in Belgium. Malvern, Ashbourne, and Highland Spring are some of the British waterways.

Minibar

A tiny refrigerator in a hotel room containing drinks that must be paid for when the hotel is checked out.

Minibus

A small bus that seats around ten people is frequently used to transport passengers from an airport to a hotel or vice versa.

Mint

Herb that is widely used Mentha is a flavoring used in cooking and to flavor commercially produced desserts.

Comment: Mint is commonly used to flavor vegetables in British food, such as new potatoes or boiling peas. It's also used as an iced drink garnish. It's commonly used in lamb dishes like keftedes in Mediterranean cuisine.

Minimum-Rated Package Tour

Inclusive air trip with only the bare minimum of accommodations to comply with regulations and qualify for tour-basing fares. In the United States, this is referred to as a "minimum land package."

Minor Operated Departments (MOD)

Phones, guest laundry and valeting, casino, and leisure club operations under direct management, as opposed to major managed divisions associated with rooms, food, and beverages, as well as rentals and concessions [Uniform System of Accounts for Hotels].

Miscellaneous Charges Order (MCO)

An all-purpose voucher issued by an airline and drawn on any organization willing to accept it as payment in advance for services or costs related to transportation covered by the airline ticket, such as lodging, meals, sightseeing, and transfers.

Mise En Place

- The act of arranging chairs, tables, and linen in a restaurant so that customers can sit down;
- The act of preparing ovens, pans, and other kitchen equipment in preparation for the day's cooking;
- The act of preparing the basic ingredients for sauces, such as chopping vegetables and other vegetables, so that they can be cooked. (*Note:* The term 'mise en place' is derived from the French phrase 'to put in place').

Miss

You will miss a bus, plane, or train because you will be late. If you don't want to miss the plane, you'll have to hurry. He missed the train and had to wait for the next one for thirty minutes.

Misunderstanding

A circumstance in which one person does not understand what the other is saying or doing, resulting in an argument. My tickets were misplaced due to a misunderstanding.

Mocha

- A dark-brown, strong-tasting coffee from Yemen and other Arabian Peninsula countries;
- Cuisine a drink or flavoring prepared by combining coffee and cocoa (Figure 57).

Figure 57. Mocha coffee.

Source: Image by Flickr.

Modem

A modem is a device that connects a computer to telephones in order to send data. To connect to the Internet, you'll need one.

Monastery

A collection of structures where monks dwell or used to live. The historic monastery has been extensively renovated and transformed into a five-star hotel. Some Greek monasteries are off-limits to female visitors.

Money

Coins and notes used for buying and selling to make money to get a wage to lose money to make a loss, not to earn a profit the hotel has been losing money for months the hotel has been working at a loss to get your money back to get a reimbursement of money which you have invested out to make money to make a profit to put money down to pay cash, particularly as a deposit. He put down a deposit of £25 and paid the remainder in installments. They are quite valuable and are worth a lot of money.

Monopoly

To have the monopoly of alcohol sales or to have the alcohol monopoly, a scenario in which one person or company controls the whole market in the supply of a product. The hotel has a monopoly because it is the only one in the town. The corporation has a total monopoly on French wine imports.

Monsoon

As a result of cyclical reversal of pressure over land and neighboring oceans, a regular periodic wind blows at specific times of the year. Most usually applied to the Indian Subcontinent and Southeast Asia, where southwest moist winds provide rains in the summer (wet monsoon), and northeast dry winds bring showers in the winter (dry monsoon) (dry monsoon).

Moonlighting

Working in one or more additional jobs outside of one's regular job, frequently on one's own dime and without declaring the income for tax purposes. With the gradual reduction in working hours in many industries and vocations, moonlighting is becoming more common in many economies. Therefore, there are numerous prospects for it in the travel, tourism, and hospitality industries, all of which offer plenty of part-time (PT) work.

Moor

Grass and tiny plants cover a wide expanse of bad terrain. The Lake District is a wild place, with moors and forests galore, to attach a boat to something ships and boats to attach a boat to something. The boat was moored to the shore of the river. He rowed his boat up to the jetty and secured it with a rope.

Montezuma's Revenge

A common phrase for diarrhea (diarrhea) caused by eating unclean fruit or drinking contaminated water while traveling overseas. Also known as Delhi belly.

More Distant Point Principle

The basic guideline that allows an airfare to be decreased in certain circumstances by including a point in an itinerary that is further away than the traveler wants to fly.

Mosquito

A little flying insect that feeds on the blood of humans and animals by biting them. A mosquito buzzed around my head, waking me up. Mosquitoes are known to carry diseases like malaria. (*Note:* Mosquitoes is the plural form).

Motion Sickness

Car sickness is a type of motion sickness produced by the movement of a vehicle, which expresses itself as abdominal pain, nausea, and vomiting. It's also known as travel sickness.

Motivations

Psychological triggers that motivate or stimulate people to behave in a certain way. As a result, whereas determinants describe the elements that support tourism growth, motives explain why people want to visit. Moreover, motivational research, or the study of the psychological factors that influence human behavior, especially in purchasing situations, can help with travel, tourism, and hospitality planning and development, as well as marketing.

Motorway

A large multi-lane road with up and down lanes split by a central strip, with limited access and vehicle restrictions due to the lack of surface intersections. By the year 2000, the network in the United Kingdom (UK) had grown to 3,316 km (2,060 miles), with other trunk roads covering 12,150 km (7,750 miles).

Motel

A hotel designed specifically for vehicle drivers, featuring unique parking spaces next to the rooms. Last Saturday, they booked into the motel. The hotel is filled, but there is a motel near the highway interchange just outside of town. Also known as a motor hotel, a motor inn, or a motor lodge.

Comment: Motels can be located on major highways and on the fringes of cities.

They often have pleasant rooms and, sometimes, a small restaurant. Larger motels may feature pools and other amenities, although they are often only used for one-night stays.

Motoring Organization

An organization that represents motorists' interests and provides services to them.

Comment: Motoring organizations provide their members with breakdown services as well as a variety of additional services such as insurance, travel information, and so on.

The AA, RAC, and RSAC are the main motoring organizations in the United Kingdom (UK); the ADAC is the primary motoring organization in Germany; and the AAA is the main driving organization in the United States.

Mountain

A phrase that serves as a descriptor but has no clear definition. An elevated location rising steeply above the surrounding landscape, usually at least 2,000 feet (600 meters), and occurring both as isolated features, such as Mt Egmont on New Zealand's North Island, and as mountain ranges, such as the Himalayas. Mountains are the principal feature of some of the world's most beautiful scenery, frequently protected from development and a popular climbing and skiing destination. Moreover, marked by vegetation zonation as climate varies with elevation and prone to fast weather fluctuations for which visitors are sometimes unprepared; such zonation and changeability are especially noticeable on solitary summits in oceanic climatic areas.

Mountaineering

Mountain climbing is a popular sport in Alpine countries, where it started, but it is now also popular in the Himalayas, Andes, and Rocky Mountains (Figure 58).

Figure 58. Mountain climbing.

Source: Image by Pixabay.

Multicultural

When referring to communities or civilizations made up of people of different races or cultures, such as those found in Hawaii and Singapore with their mixed populations.

Multilateral

Concerning more than two sides or parties, such as in multilateral accords involving more than two countries, with no distinction made between those engaged. Multilateral accords address several parts of international travel and tourism that are regulated, such as transportation and communication between countries.

Multinational Company/Corporation

An international firm, often known as a transnational corporation, is a significant business enterprise that operates in several nations. The majority of examples in travel, tourism, and hospitality may be found in the hotel industry (e.g., Hilton International, InterContinental Hotels Corporation, Sheraton), as well as vehicle rental (e.g., Avis, Eurodollar, Hertz) and other areas.

Multiple

The term is most usually used to describe a shop or store that is similar in nature and operated by the same people in different places. Following their fast growth, the plural form, multiples, became widely used in the UK in the 1980s to refer to travel agency firms with branch offices.

Multiple Land Use

Land use planning refers to the use of land for multiple purposes, one of which is commonly recreation (both private and public). Typical of low-intensity usage zones, such as highland grazed by domestic animals and used for field sports, water collecting, and informal recreation. Even if two uses are incompatible, they can coexist if their seasons do not coincide. Shooting and hunting are also permitted in areas of intense farming.

Municipal

Of or relating to local government/authorities. Many travel and tourism enterprises, such as museums, galleries, and other tourist/visitor attractions, as well as many public passenger transportations, are held by municipalities in the United Kingdom (UK).

Murphy

When not in use, an American phrase for a bed that folds or swings into a cabinet or wall.

Museum

An organization that collects, preserves, displays, and interprets exhibits, primarily in the fields of art, history, religion, and natural history. Most fields of human endeavor are now represented in museums, many of which are popular tourist destinations. As a result, there are museums dedicated to a variety of sports, hobbies, and other interests.

N

Nansen Passport

Fridtjob Nansen, a Norwegian statesman and humanitarian, was the inspiration for this internationally recognized refugee identity certificate.

National Estate

The Australian Heritage Act 1975 defines heritage as "those areas within Australia's cultural or natural environment that have esthetic, historic, scientific, or social significance or other unique worth for future generations as well as for the current community." These places are generally prominent tourist/visitor attractions and are placed on the Register of the National Estate.

National Holiday (Vacation)/Travel/Tourism) Surveys

In some circumstances, sample enquiries of homes or people are used to determine the amount, value, and features of holiday (vacation) journeys/visits, whereas in others, all or most excursions are measured. Most Western

European countries conduct annual surveys, some of which are repeated; similar polls are also conducted in Australia, the United States, and other countries. Moreover, the OECD Tourism Committee has issued guidelines for national holiday surveys in order to facilitate survey comparability, however these are implemented to differing degrees in different countries.

National Nature Reserves

More than 300 areas in the UK are managed by English Nature, Scottish Natural Heritage, Countryside Council for Wales, and the Department of the Environment for Northern Ireland in order to conserve natural habitats and their plant and animal species, as well as to provide opportunities for research with controlled access. Other plans can be found under rural conservation designation initiatives.

National Parks

Large rural areas, sometimes owned by the government, are declared as such by the government to safeguard and retain their natural beauty and other aspects. The idea began with the establishment of Yellowstone National Park in 1872, and therefore US National Park Service currently oversees over 300 parks, historic sites, and recreation areas.

However, in Great Britain (GB), the 15 National Parks that exist or are scheduled to be designated by 2002 (10 in England, 3 in Wales, and 2 in Scotland) are primarily in private ownership, and recreational use coexists with agricultural production and other activities on land with public access, within a framework of building development and other controls (Figure 59).

Figure 59. Manas National Park, Assam, India.

Source: Image by Wikimedia Commons.

National Readership Survey

The Joint Industry Committee for National Readership Surveys (JICNARS) conducted a continuous sample study of more than 50,000 people in the United Kingdom (UK). It therefore examines the average issue readership of major periodicals, as well as ITV viewing, commercial radio listening, movie attendance, and specific interest groups such as car-owning households. Hence, JICNARS summarizes the findings, and Social Trends publishes a summary of the findings.

National Scenic Areas (Scotland)

In Scotland, forty tracts of mostly privately held property have been declared in order to preserve their scenic beauty through stringent development control, but with no defined recreational use. They cover more than one-eighth of Scotland's land area, similar to areas of outstanding natural beauty (AONBs) in other parts of the UK.

National Tourism Administration (NTA)

The World Tourism Organization (WTO) defines it as:

- A central government body with top-level administrative responsibility for tourism;
- A central government body with direct intervention powers in the tourism industry;
- All national government administrative authorities with the authority to interfere in the tourism sector.

Other lower-ranking governmental or official bodies – whether incorporated into a larger body or operating independently – may be considered NTA executive bodies. Therefore, central entities that are legally or financially tied to the NTA may also fall under this category.

National Tourism Organization (NTO)

Additionally, alluded to as National Tourism (or Tourist) Office, an authority body differently worried about the turn of events, advancement, and coordination of the travel industry in a nation, perceived, and furthermore to a changing degree funded by its administration. Moreover, three principal types are legislative (government division), semi-administrative (legal body set up by the government), and nongovernmental (deliberate affiliation).

Hence, the assignment might vary as needs be, e.g., Ministry of Tourism, Tourist Board, Tourist Association.

National Tourist Boards (UK)

Legal bodies set up under the Development of Tourism Act 1969 are liable for the travel industry advancement in their particular nations and for their advancement inside the UK: English Tourist Board (ETB), presently English Tourism Council (ETC), Scottish Tourist Board (STB), presently VisitScotland, Wales Tourist Board (WTB). Therefore, these were gone before by the Northern Ireland Tourist Board (NITB) laid out under the Development of Tourist Traffic Act (Northern Ireland) 1948.

National Travel/Tourism

From the place of a nation, includes for measurable purposes homegrown travel/the travel industry in addition to outbound travel/the travel industry, e.g., French inhabitants' movement/the travel industry in France and abroad.

National Travel Survey

English nonstop example overview done for the Department for Transport, Local Government, and the Regions by Office for National Statistics, which gathers information on private travel designs in Great Britain (GB) and relates various types of individual travel with the qualities of explorers and their families. Further, discoveries are distributed by the Department of Transport Statistics in GB. To be recognized by British National Travel Survey (BNTS).

National Vocational Qualifications (NVQS)

A complete system of professional capabilities for all occupations and callings in England, Wales, and Northern Ireland is planned basically for youngsters between the ages of 16 and 19 who have left full-time schooling, giving position explicit abilities and information. Therefore, there are five levels going from establishment to expert and the board and the areas incorporate, i.e., expressions, and diversion, catering, and accommodation, natural preservation, game, and amusement, travel administrations, warding, and guest administrations. Hence, the Scottish identical to NVQs is Scottish vocational qualifications (SVQs).

Nationality

The ethnicity of a voyager is that of the Government giving his/her visa (or other recognizable proof records), regardless of whether the person typically dwells in another country [World Tourism Organization (WTO)].

Natural Heritage

Normal elements comprising physical and organic arrangements of extraordinary widespread worth from the tasteful or logical perspective; topographical or geographical developments and living spaces of undermined types of creatures and plants of exceptional all-inclusive worth according to the perspective of science or preservation; regular locales or areas of remarkable general worth according to the perspective of science, protection or of normal excellence [based on UNESCO Convention for the Protection of the Cultural and Natural Heritage, 1972] (Figure 60).

Figure 60. Natural heritage – Bhutan.

Source: Image by Wikimedia Commons.

Nature Reserve

An area of land or water is put away to be so overseen as to secure and moderate its creature and vegetation and different elements and is frequently utilized for research. In the UK there are both nearby and National Nature Reserves, some claimed and oversaw openly, some secretly, and others by intentional bodies, like the Royal Society for the Protection of Birds.

Nature Trail

A way through a recreation area or the field with signs to cause to notice significant and fascinating highlights, like plants, trees, birds, or creatures.

Naturism

Equivalent word for nudism, confidence in and practice of going about stripped. Consequently naturist (an individual having confidence in and working on going about exposed); naturist ocean side (ocean side where the training is permitted); naturist settlement (club or community for naturists).

Neap Tide

A tide which happens at the first and last quarters of the moon, when the distinction between high and low water is not exactly ordinary.

Neoclassical

Style of engineering and plan of the late eighteenth and mid-nineteenth hundreds of years addressed a re-visitation of traditional Greek and Roman structures as a response against Baroque and Rococo.

Neo-Colonialism

The term is used to depict the impact a strong nation practices over the undertakings of another, generally over monetary and political issues of a less evolved country. This might incorporate specialized and improvement help and the exercises of global organizations in the movement: the travel industry and accommodation.

Neutral Unit of Construction (NUC)

The premise of air charge estimations was laid out by the International Air Transport Association (IATA), which supplanted the fare construction unit

(FCU) on July 1st, 1989. In the new framework, individual sums in nearby monetary standards are changed over completely to NUC by the utilization of NUC transformation factors and the all-out NUC sum is switched back over completely to the neighborhood money of offer. Therefore, the change factors are regularly changed four times each year.

No Show

The term is utilized in the business for a lodging visitor or traveler with a booking who neglects to take it up without advising the inn or transporter.

Nomad

An individual from a race or clan moves from one spot to another to observe the field, subsequently, one who carries on with a meandering life. Therefore, wanderers are typically barred from movement and the travel industry insights (Figure 61).

Figure 61. Nomad camp after sunset.

Source: Image by Flickr.

Non-Governmental Organization (NGO)

Worldwide association laid out by people, affiliations, or firms (i.e., not by between administrative understanding), in which individuals will be people or corporate bodies. Therefore, the association makes its own resolutions, can't guarantee an advantaged position, and is regularly dependent upon the law of the state where its central command or secretariat is found. European Travel Commission (ETC), International Hotel and Restaurant Association (IHRA) and Pacific Asia Travel Association (PATA) are notable models.

Non-Price Competition

A conventional term depicting factors other than cost, which separate the results of one vender from those of another and impact the purchasers' decision. Therefore, there is a lot of proof to recommend that providers of numerous labor and products like to contend on factors other than cost and this is likewise progressively the situation in movement, the travel industry, and friendliness. For such offices as lodgings, the area is a typical cutthroat variable. Moreover, reservation administrations comprise a serious comfort for purchasers of numerous vacationer items, as do credit courses of action, which empower records to be settled without the installment of money.

North American Free Trade Agreement (NAFTA)

An arrangement between Canada, Mexico, and the USA, which became effective on 1 January 1994, started a steady course of taking out taxes and different boundaries to exchange between the three nations.

While it doesn't manage travel, the travel industry and cordiality, they are covered all the more for the most part under such headings as exchange administrations, speculation, and brief section, as well as under ecological and work issues. However, critically, the travel industry is supposed to develop because of the fortified economies of the nations and NAFTA invigorates expanded participation between their National Tourism Administrations (NTAs).

Northern Lights (Aurora Borealis)

A moving hued shine is seen close to the skyline in the night sky in high elevations in northern pieces of Scandinavia and somewhere else in the northern half of the globe. The comparing peculiarity in the southern half of the globe is aurora australis (Figure 62).

Figure 62. Northern lights – Aurora Borealis Norway.

Source: Image by Wikimedia Commons.

Northern Summer

The term is utilized in Australia and New Zealand for the late spring season in the northern half of the globe (typically alluding to the long stretches of June, July, and August), which concurs with winter in Australia and New Zealand.

Nostalgic Trains

US expression for trains reestablished to the previous state and worked as extraordinary visits and vacation destinations as opposed to a method for transport, for example, Orient Express.

Notice of Proposed Rule Making

A public declaration by an administration office, for example, a common aeronautics authority, that it is thinking about correcting its current guidelines, giving a chance to closely involved individuals to spread the word.

Notifiable Disease

Genuine irresistible sickness, for example, cholera, meningitis, or typhus, which must be accounted for by a specialist to the proper power (the Department of Health on account of Great Britain (GB)), from whom the full momentum rundown of notifiable infections is accessible.

O

Oats

A grain food, filled in northern European nations.

Comment: Oats are broadly developed and utilized in Scotland, where the most well-known use for them is in making porridge and rolls.

Occupancy

The rate relationship of convenience limit used to the accessible limit in inns and comparative foundations, a proportion of usage closely resembling load factor in transport. Bed/visitor/sleeper inhabitance relates involved beds to accessible beds, and room inhabitance relates consumed rooms to accessible rooms. Henceforth single/twofold inhabitance means room inhabitance by one/two people.

Occupation

The control of any individual is the sort of work the person performs, due respect being paid to the circumstances under which it is played out; this by itself decides the specific gathering in a word-related characterization to which the individual is doled out. Therefore, the idea of the processing plant, business, or association in which the individual is utilized has no bearing upon the order of his/her occupation. Then again, it might empower the idea of his/her obligations to be all the more obviously characterized.

Oceania

A term used to signify: (i) the Pacific Islands and the adjoining oceans; or (ii) the district including Australia, New Zealand, and the Pacific Islands.

Off-License

A shop auctioning liquor for utilization off the premises, likewise the name for one of two fundamental licenses available to be purchased of alcoholic alcohol in England and Wales, approving offer of all classes of alcohol yet just for utilization of the premises for which the permit is allowed. The permit is therefore conceded by permitting judges working in each authorizing locale. The Scottish identical is off-deal permit; the US same, bundle store.

Off-Season

Occurring during the season when fewer individuals are voyaging or utilizing offices thing TOURISM the season when there are fewer explorers, thus admissions and room costs are less expensive. Tour administrators ask more individuals to go in the slow time of year. Therefore, air tickets are less expensive during the slow time of year. Additionally, called low season.

Oil

A thick smooth-running fluid of different sorts used in cooking action word to put oil on Oil the tin prior to placing the batter in. Remark: The commonest cooking oil is produced using sunflower seeds; numerous others exist, for example, olive oil, pecan oil, and so forth.

Okra

A vegetable with a green unit is utilized in soups. Likewise called gumbo 1, woman's finger.

Comment: Also called 'bhindi,' 'gumbo' or 'woman's finger.' Utilized in Caribbean and Indian cooking; additionally, utilized in the south of the USA (Figure 63).

Figure 63. Okra (Abelmoschus esculentus), known in many English-speaking countries as ladies' fingers or Ochro.

Source: Image by Wikimedia Commons.

Olive

A little dark or green natural product from a Mediterranean tree, which is squashed to create oil and is additionally eaten as food.

Omelet

A dish made of beaten eggs, cooked in a griddle, and collapsed over prior to serving. Different fillings might be added.

Comment: The commonest types of Omelets are cheddar omelet and ham omelet. A Spanish omelet is made with onion, tomato, peppers, potatoes, and so on, and isn't collapsed over, yet all at once served level.

Old World

Ordinarily comprehended to indicate the eastern half of the globe, particularly Europe, Africa, and Asia eastwards to the extent that the Malay Archipelago.

Oligopoly

In financial matters, a type of flawed contest exists when there are only a few vendors on the lookout. Assuming their items are homogenous – each an ideal substitute for the others – the oligopoly is portrayed as awesome; in the event that the items are separated, it is a flawed oligopoly. Therefore, the previous is seldom found practically speaking. However numerous convenience and cooking markets will generally fit the depiction of blemished oligopoly. E.g., in numerous towns, a couple of inns address the greater part of the convenience limit of the area, and proposition comparable offices at comparable costs.

Olympic Games

A quadrennial worldwide game gathering held in different spots (the first in quite a while in Athens in 1896) and a significant vacationer/guest fascination. The principal games after World War II were held in London in 1948, then, at that point, Helsinki (1952); Melbourne (1956); Rome (1960); Tokyo (1964); Mexico City (1968); Munich (1972); Montreal (1976); Moscow (1980); Los Angeles (1984); Seoul (1988); Barcelona (1992); Atlanta (1996); Sydney (2000); the 2004 Games are planned to be held in Athens.

Omnibus Survey

Ordinary recurrent overview utilizing a normalized system and covering various themes for various clients simultaneously. Individual clients commission a predetermined number of inquiries each from a statistical surveying organization working such reviews from month to month or other customary spans. This, is essentially more affordable for them than reason planned separate overviews and an especially practical methodology for much statistical surveying in movement, the travel industry, and friendliness, however, it gives restricted extension to itemized discoveries.

One-World Worldwide

Carrier coalition of (June 2001) Aer Lingus, American Airlines, British Airways, Cathay Pacific, Finnair, Iberia, Lan Chile, and Qantas, shaped February 1999 with American, British Airways, Qantas, and Cathay Pacific as establishing accomplices.

On-License

One of two primary licenses for the offer of alcoholic alcohol in England and Wales, which allows the offer of all or a few glasses of alcohol for utilization in the vicinity and on-permit. One of two fundamental licenses for the offer of alcoholic alcohol in England and Wales, which allows the offer of all or a few glasses of alcohol for utilization anywhere nearby and furthermore endorses deal for utilization of the premises for which the permit is conceded. Therefore, the permit is allowed by permitting judges working in each authorizing region.

On-Line/Off-Line

- In transport by and large, nations or spots through which a transporter works with freedoms to get travelers are portrayed as online; those through which a transporter does not work or does not have privileges to get travelers as disconnected.
- In air transport, utilizing a similar aircraft, e.g., changing planes with a similar carrier, is portrayed as an online association, in particular from changing planes as well as aircraft, depicted as a disconnected association.
- In figuring, associated with and speaking with a PC as, e.g., travel service outfitted with a terminal connected to a transporter's PC is online.

On-Airport Car Rental Firm

A vehicle rental firm with its base inside an air terminal complex, is more advantageous for voyagers, albeit the rates might be higher.

On-Your-Own Package

Plans are made by a travel service for an autonomous voyager who just needs the aircraft reservation and an inn.

Open Country

An enlightening term for crude land covered with non-tree vegetation, including grasses and bushes, especially in upland and mountain regions. England and Wales gave legitimate definition in the National Parks and Access to the Countryside Act (1949) as 'mountain, moor, heath, down, precipice or foreshore,' however hence stretched out to incorporate forest.

Open-Jaw Game Plan

- A framework where a traveler flies to one air terminal on the outward flight and gets back from another air terminal, having gone between the two by transport, train, vehicle, and so forth;
- A framework where a traveler leaves from one air terminal on the outward excursion and gets back to one more on the return excursion, or flies to one air terminal and afterwards gets back from another air terminal.

Open Rate

Air transport term depicts what is happening which exists when meetings, states, or transporters neglect to concur on uniform rates for specific courses, leaving the gatherings worried to concur their own admissions.

Open Return

A return (full circle) ticket gave by a transporter without a booking to utilize specific assistance on the bring venture back.

Open Skies

The idea of an unregulated economy for worldwide aeronautics, which permits aircraft to fly anyplace without limitations, e.g., American and British carriers between the two countries. Notwithstanding, existing open skies arrangements closed independently between part conditions of the European Union (EU) with the USA pointed toward changing air traffic between the USA and a solitary European nation have been tested by the European Commission as oppressing other part states and acting against deregulation inside the Union.

Open (-Date) Ticket

A travelers' ticket is given by a transporter, typically an aircraft, without a booking to utilize specific assistance, such reservation to be made sometime in the not-too-distant future.

Open-Jaw Trip

A return (full circle) venture with various starting and ending focuses (e.g., London/New York/Paris) or with a flight point for the return, is not the same as the appearance point (e.g., Chicago/London and Paris/Chicago).

Operational/Operations Research (OR)

A multidisciplinary way to deal with the arrangement of quantifiable business or regulatory issues, utilizing numerical cycles, for the most part with the guide of PCs.

Opportunity Cost

The expense of an open door is inescapable as, e.g., the deficiency of income that would be procured from an elective utilization of the amount of cash. This is a significant idea of commonsense worth to be applied in venture choices when a decision must be earnestly committed to the responsibility of scant capital assets to various activities.

Option

- Movement, things, or administrations are excluded from the essential value, which might be picked by the client for an extra charge, e.g., touring, or a solitary room on a visit, at which costs depend on two people sharing a room.
- A booking is made and held by an administrator, like a transporter, inn or visit administrator for a restricted timeframe, toward the finish of which the reservation should be affirmed or it is dropped.

Ordnance Survey

Previously, a British government official, presently a public organization, which is answerable for delivering definite guides of the UK. We utilized an Ordnance Survey guide to plot our course.

Oregano

Atypical spice, used in Italian dishes like pizzas (Figure 64).

Figure 64. Oregano spice.

Source: Image by Wikimedia Commons.

Organic Food

Food developed on ranches that don't utilize regular agrichemicals; however normal manures and regular types of bug control. Ranchers wishing to meet all requirements for perceived natural marks should adjust to a severe arrangement of rules. Therefore, utilizing natural products is an interesting selling suggestion of certain cafés.

Organic Farming

A technique for cultivating doesn't include utilizing compound manures or pesticides.

Organization Chart

A chart of individuals working in different divisions, showing how an organization or office is coordinated.

Orient

The nations east of the Mediterranean or of Southern Europe, including differently those of South-West Asia, the far east (FE) or Asia by and large, the vast majority of them with much potential for the travel industry advancement.

Orienteering

The game of observing your direction through harsh territory in which you don't have the foggiest idea, utilizing guides and compasses the skis with the orienteering crew.

Orient Express

Named train administration initiated 1883, shut down 1977 and revived 1982 as the Venice Simplon-Orient Express, among London and Venice through Paris and the Alps, going by Pullman among London and Folkestone, Sea Cat to Boulogne, and thereupon in refurnished unique carriages to Venice.

Origin

In air transport, the beginning stage of a travelers' venture. Whenever an excursion includes more than one leg, fragment or area utilizing more than one carrier, the aircraft conveying a traveler on the primary piece of the excursion is known as the beginning carrier.

Origin Country

For motivations behind worldwide the travel industry measurements, the nation where the guest has resided for the majority of the past a year or has resided for a more limited period and means to return in somewhere around a year [World Tourism Organization (WTO)].

Original

- New and different She created a few exceptionally unique thoughts for the plan of the new town place;
- That isn't a duplicate I kept the first receipt and sent a duplicate;
- To start with, or made interestingly bath is the home of the first Bath Oliver.

Outback

The term utilized for somewhat remote and scantily populated areas of Australia is called past the dark stump and, furthermore, Back o' Bourke in Australian English.

Outdoor Advertising

Promoting medium comprising of banners, enlightened signs, and open-air shows. Promoting locales in railroad and underground stations and publicizing spaces in transports, trains, and different vehicles are depicted independently as transport publicizing.

Outsourcing

The purchasing of items and administrations from outside providers as opposed to making them accessible inside. Normal reasons incorporate the absence of mastery, venture capital or actual space, as well as the capacity to purchase more economically or more rapidly than providing in-house.

Out of Season

Food is more costly on the grounds that the developing season is finished and it must be imported, or which is not found on menus in light of the fact that the ideal opportunity for it is over to develop it.

Oven

An encased box which can be warmed for cooking. Cook the vegetables in a microwave (Figure 65).

Figure 65. Microwave oven.

Source: Image by Pixabay.

Overbooking

Reservation of additional seats by a vehicle transporter or rooms by an inn than are accessible, some of the time-conscious to make up for expected flake-outs. In North America, likewise called oversell.

Overflights of Antarctica

Direct trips of 10–12 hours on board QANTAS Boeing 747 airplanes were withdrawing from Australian air terminals and overflying the Australian, New Zealand, and French Antarctic regions. The 'flightseeing' contract trips burn through 3–4 hours over Antarctica however include no arrivals on the landmass.

Overriding Commission

Extra commission paid ahead (like an aircraft or visit administrator) to a travel planner in specific conditions, e.g., as a little something extra or motivator. The term is likewise used to portray the commission permitted by a head to a general sales agent (GSA) in regard to appointments started by a normal travel planner, the GSA going about as a broker between the head and the travel planner.

Overtime

Hours worked more than the typical working time Overtime is paid at one and a half times the standard rate. Modifier working more than the typical working day Overtime pay is determined at one and a half times the standard rate. They stayed at work longer than required when the inn was full.

Overtrading

A circumstance emerges when a firm builds its result and deals without having satisfactory extra assets accessible to fund the extra working capital required.

Ozone Layer Depletion

Exhaustion of a slight layer of oxygen-related gases that circles the earth and channels hurtful disease-causing bright beams from the sun. Therefore, caused essentially by made gases, the reasonable impacts of the consumption are such wellbeing risks as the expanded rate of skin disease and eye harm, which could prompt a decrease in customary ocean side occasions and many water sports, as well as mountain-based the travel industry.

P

Pacific Rim

The general name was given to nations lining the Pacific Ocean, especially on the northern side of the equator, where the volume of exchange and vacationer development, outstandingly between East Asia and North America and inside the locale, is developing quickly.

Package

For the most part, a blend of at least two components is sold as a solitary item at a comprehensive cost, in which the expenses of the singular item parts are not independently recognizable. In the movement and the travel industry, the term is utilized as a well-known equivalent word for a comprehensive visit, as in bundle occasions (get-aways) or bundle visits.

Package Store

American expression for an alcohol store auctioning liquor for utilization of the premises is comparable to an off-permit in England and Wales.

Package Travel Directive European Community (EC)

Chamber Directive to inexact the regulations and guidelines of part states connecting with the deal, or make available for purchase, of bundle occasions, which was concurred by EC Consumer Affairs Ministers in June 1990 and came into procedure on 1 February 1993. Moreover, the Directive characterizes the terms bundle, travel coordinator and travel retailer, and sets down necessities for the security of the customer. Hence, the Department of Trade and Industry (DTI), answerable for carrying out the Directive in the UK, created the Package Travel, Package Holidays and Package Tours Regulations 1992, concurred by Parliament in December 1992.

Parador

Lodging in Spain, one of a chain possessed and constrained by the state, giving an elevated expectation of the travel industry convenience in conventional Spanish style, frequently changed over from a palace, cloister or another noteworthy structure (Figure 66).

Figure 66. Parador Alhambra Granada, Spain.

Source: Image by Wikimedia Commons.

Parallel of Latitude

A line drawn on a guide connecting all focuses on the world's surface with a similar precise separation from the equator and, thusly, circling the earth corresponding to the equator. There are 90 divisions or degrees between the equator and every one of the shafts, every degree is partitioned into an hour and every moment into 60 seconds.

Park-and-Ride

A plan which empowers drivers to leave their vehicles and proceed with their excursion by open vehicle progressively used to lessen gridlock in metropolitan regions and in delicate country regions.

Pars

US PC reservation framework (CRS) claimed by Trans World Airlines (TWA) and Northwest Airlines, converged with DATA II and both were supplanted by the World span framework in 1990.

Participation Rate

The extent of a section in a specific populace action, like open-air entertainment or the travel industry, as particular from movement rate, which alludes to business.

Partnership

Type of responsibility for carried on by at least two individuals in like manner, sharing dangers, benefits, and misfortunes. Regularly each accomplice is liable for the obligations and for the activities of others to the full degree of their own belongings. This is a typical type of proprietorship in the callings and in exercises with generally little capital prerequisites, as in some cooking and retailing. Moreover, in numerous nations, associations are represented by regulation, e.g., in the UK by the Partnership Act 1890. Association of reciprocal gatherings, e.g., aircraft, and vehicle employ (rental) or public and private areas.

Part-Time (PT)

By and large, more limited hours than establishing a full working week in a specific industry or occupation. Nonetheless, what is part-time (PT) business is typically characterized for specific purposes; subsequently, e.g., temporary laborers are characterized for measurable purposes by the UK Department for Education and Skills as individuals regularly working for not over 30 hours per week. Travel, the travel industry and cordiality exercises are enormous bosses of seasonal work.

Passenger Designations

- **Through Traveler:** One planned to go on a similar vehicle to its last objective.
- **Transfer Traveler:** One who changes vehicle to proceed with an excursion by associating administration.
- **Transit Traveler:** One who breaks an excursion at a middle of the road point, to change the vehicle or for another explanation; if going between nations, not leaving the travel region of the air terminal or port and, in this way, not authoritatively entering a country; such a traveler isn't dependent upon passage conventions, for example, customs control.

Passenger Space Ratio (PSR)

Otherwise called transport thickness, the proportion of gross enlisted weight (GRT) and traveler limit, is of specific importance in voyage ships: low thickness sends ordinarily give more space in lodges, public rooms and on decks, as well as the other way around.

Passport

An authority report was given by an administration to a resident of a nation, checking his/her character and citizenship, for moving abroad.

Payload

The income delivering heap of an airplane, including the two travelers and freight. It approaches the all-out drop weightless the vacant weight including hardware and working burden (fuel, supplies, and flight group). Moreover, the dependable government specialists fix the greatest take-off payload limit of every airplane.

Pension Establishment

Normal in Continental Europe, generally depicted in English-talking nations as a visitor house; ordinarily, the proprietor oversaw and gave convenience, food, and drink to occupants as it were.

Per Capita

Deeply grounded, albeit wrong, a term utilized rather than the right per caput, signifying 'a head' or 'each,' as, e.g., per capita pay (normal pay per head), and per capita use (normal consumption per head), both normal measurements got from customer market studies, including guest overviews.

Per Diem

Installment is made as a proper sum each day by firms and different associations to workers and others, for example, specialists, who invest energy away from their base, to cover their everyday costs.

Perfect Competition

In financial aspects, a market circumstance with an enormous number of vendors of a homogeneous item. Every merchant supplies just a little part of the complete result, so no dealer can alone influence the market cost. Therefore, every dealer supplies an item indistinguishable from the results of the relative multitude of different vendors, for which it is an ideal substitute, so purchasers can have no inclination between the results of various merchants. Hence, not many business sectors in movement, the travel industry and cordiality or somewhere else even methodology amazing contest, however, the model is useful for understanding business sectors by and large and different types of flawed rivalry specifically.

Personal Income

Pay got by people from work, ventures, and move installments as wages, compensations, other work income, profits, interest, and lease.

Personal Selling

Likewise alluded to informally as eye-to-eye selling, the term portrays individual contact with oral introductions to possible purchasers.

Albeit such structures as deals calls by delegates to buyers are not normal in movement, the travel industry and accommodation, they are utilized in arranging bunch deals, likewise with gathering and presentation coordinators, and by makers versus go-betweens as, e.g., by visit administrators opposite retail travel planners.

Personnel Association

A relationship of representatives in a specific occupation to advance their normal advantages which, contingent upon its extension and capacities, may yet require not be an expert body or a worker's organization. Moreover, models incorporate the Association of Conference Executives (ACE), and the Association of National Tourist Office Representatives (ANTOR).

Person-Trip

A standard proportion of the travel industry movement, recording every individual making an excursion; for instance, three people making an outing, rises to three-man trips.

Phare

Abbreviation initially shaped from the French expression for Poland and Hungary; Assistance for Economic Restructuring, a plan sent off by the European Community (EC) in 1989 to assist the change with handling, at first in Hungary and Poland. In this way other Central and Eastern European Countries (CEEC), including the Baltic States, have likewise benefited. Projects of specialized help have stretched out to the travel industry improvement.

Physical Planning

The method involved getting ready recommendations for and directing the utilization of land in a given region. Metropolitan (town) arranging alludes to actual preparation in a developed region, rustic (country) wanting actual preparation outside a developed region. Therefore, the term is now and again utilized related to advancement while building and different activities or changing the utilization of land or structures are involved. Moreover, arranging and improvement, along with showcasing, are the chief capacities in the travel industry, and this is reflected in the extension and construction of public, provincial, and nearby travel industry associations.

Pickup Point

Place where people or gatherings are gathered by a mentor or one more vehicle by an earlier course of action. (b) Slang expression for a spot where whores assemble to meet clients.

Pidgin-English

Communicated in language comprising to a great extent of English words, frequently undermined in pronunciation, initially primarily utilized for correspondence between the Chinese and Europeans, presently all through the Pacific Islands where it is the most normal most widely used language between different neighborhood bunches communicating in numerous nearby dialects.

Pilgrimage

Development of devotees too far off heavenly places. The most inescapable of these is likely the excursion to Mecca, which all Muslims should make no less than once during their lifetime. Previously made by walking, by camel or by transport, yet presently progressively via air, on planned administrations and by contract, from West Africa, Indonesia, and somewhere else. Hindus therefore travel to Varanasi (Benares) in India, Japanese to tribal sanctuaries all through the nation, and Christians and Jews to Jerusalem. Moreover, visits to strict locales from the created world are currently bound to be made in light of the verifiable or structural interest of the sacred places as opposed to basically or solely for strict purposes. In any case, Lourdes in France and Knock in Ireland are instances of present-day altars that have emerged from dreams and draw in enormous quantities of strict guests (Figure 67).

Figure 67. A pilgrimage in India.

Source: Image by Flickr.

Pink Dollar

A term used to allude to spending by gay and lesbian purchasers who have arisen as a critical and recognizable traveler market fragment, in certain nations served only by experts in the travel industry and friendliness administrators.

Planetarium

A functioning model of the planetary framework, typically encased in a domed structure, is a critical traveler/guest fascination in numerous urban communities.

Point of Sale (POS)

Where selling happens. For most travel, the travel industry and accommodation items outside retail locations incorporate retail travel services and administrators' reserving workplaces; inner or in-house retail locations incorporate, e.g., inn gathering workplaces, bars, and cafés. Therefore, show and other advancement material utilized there is additionally alluded to as retail location.

Polder

An area of low-lying land at or beneath ocean level recovered from the ocean, lake or waterway by dykes and depleting; normal, e.g., in the Netherlands.

Polio (Myelitis)

A viral illness of the sensory system, causing muscle torments, fever, and loss of motion; can cause extremely durable handicap and is in some cases deadly. It is therefore spread by waste oral means and by bead disease (hacks and sniffles) in areas of poor hygiene (outside North and Western Europe, North America, Australia, and New Zealand). Might be forestalled by vaccination, which must be trailed by a promoter portion at regular intervals.

Pollution

Immediate or aberrant adjustment of the properties of any piece of the common habitat so as to introduce a peril or expected danger; on the other hand, presence of material in an unlucky spot and in some unacceptable structure. Contamination can hence influence the climate through destructive gases or particulate matter, the land through the unloading of waste or poisonous matter, and the ocean, waterways, and lakes through releases or testimony.

Nevertheless, vacationers have been both unfavorably impacted by contamination, e.g., by traffic in urban communities, by the defilement of water supply and the release of sewage on washing seashores, and been a wellspring of contamination, particularly where numbers are huge and the travel industry arranging is lacking.

Polynesian Bure

Covered, cabin style convenience is found all through the South Pacific and in island resorts of Australia in town style resort settings. Albeit the term has turned into a practically nonexclusive articulation, phrasing specifically nations differs: e.g., a similar convenience is referred to in Samoa as Fale.

Pooling

In air transport, sharing the income to be acquired on a specific course between at least two carriers, on a foreordained premise by an understanding between them, which may likewise cover matters, for example, helpful flight planning and joint promoting.

Population Change

The fundamental parts of progress in the size of the number of inhabitants in space are regular change (i.e., the contrast among births and passings) and net movement (i.e., the distinction between migration and resettlement). Patterns in these are of movement and the travel industry importance in light of the fact that the size of the populace is a fundamental determinant of the volume of movement and the travel industry created by an area.

Port

Harbor or town having a harbor. (b) The left half of a boat or airplane while looking ahead, with red route lights on the scaffold or wings.

Poste Restante

The French term in worldwide use for the mail center office where mail so tended to is kept until called for by the recipient, subsequently applied to the assistance, which is of specific advantage to explorers. General Delivery is the US same.

Post-Industrial Society

The term used to depict a phase came to by society in the late 20th 100 years in North America, Western Europe, and a few different pieces of the created world, because of innovative advancement and financial and social change. Therefore, primary qualities incorporate the general decay of products delivering ventures, the development of administrations and expansions in expert and specialized occupations. Hence, the development of movement and the travel industry is a noticeable component of the post-modern culture.

Pousada

An inn in Portugal, one of an organization of decisively found foundations to address the issues of visiting drivers, giving the travel industry convenience, frequently in changed over memorable structures (Figure 68).

Figure 68. Pousada De Palmela – Portugal.

Source: Image by Wikimedia Commons.

Power Boating

Make fueled essentially by petroleum motors, both as an end in itself, whether for speed sailing or cruising or as a wellspring of footing for water-skiing and paragliding. The two structures are much of the time a wellspring of contention with different types of amusement on or in, or on the banks of, water, as well as wellsprings of disintegration of delicate banks from the wash, and of contamination from the release of oil, sewage, and other waste.

Precipitation

All types of dampness fall on the ground, whether as downpour, snow, slush, hail, or haze trickle. All structures are estimated as what could be compared to rain, one foot (approx. 30 cm) of snow being comprehensively identical to one inch (approx. 2.5 cm) of a downpour.

Preclearance

Arrangement of customs and movement methodology in a far-off nation of takeoff to facilitate the interest for such offices in the nation of appearance. Such courses of action exist, e.g., between a few Caribbean nations and the USA.

Predatory Pricing

The act of a firm briefly selling at exceptionally low costs with the target of driving out contenders or keeping out new participants from a market. When the goal has been accomplished, costs can be raised and higher benefits made.

This is the very thing Laker Airways asserted a few carriers were doing in the last part of the 1970s and mid-1980s when they cut their charges on North Atlantic courses on which Laker worked its Skytrain administration, and what Virgin Atlantic whined of all the more as of late. Moreover, the training is challenging to demonstrate however has been likewise claimed to exist, e.g., in Europe in the opposition among planned and contract aircrafts.

Prepaid Ticket Advice (PTA)

Warning by a specialist or transporter that an individual has paid for someone else's transportation, normally from a spot other than the one wherein the passage was paid, in this manner approving the issue of an aircraft ticket by the beneficiary.

Pre-Registration

Term to depict methods for the consummation of enrollment preceding the visitors' appearance, utilized for meetings, displays, visits, and other huge

relaxation, entertainment, and occasions; visiting companions and family members; business and experts; wellbeing treatment; religion/journey; other.

Push-Pull Theory

Hypothesis initially utilized regarding relocation, which recommends that individuals are moved by antagonistic circumstances (like joblessness or political restraint) to leave a region and are simultaneously drawn in by an area with great circumstances (for example, business possibilities or opportunity). Moreover, the hypothesis has been all the more as of late stretched out to make sense of by similarity to the travel industry, where it appears to be similarly undeniable (whether on climatic or different grounds).

Q

Qualifier Group

Aircraft partnership of (June 2001) Air Europe, Air Liberté, Air Littoral, AOM, Crossair, LOT Polish Airlines, PGA Portugalia, Swissair/Sabena, TAP Air Portugal, Turkish Airlines, Volare, shaped March 1998.

Quango

An abbreviation shaped from truncation of semi-independent non-legislative association, a semi-public body in the UK delegated and funded by the government, however not an administration office. Moreover, mdels incorporate the British Tourist Authority (BTA) and other legal vacationer sheets.

Quarantine

Separation is forced by well-being experts on people or creatures that could spread irresistible infection. Whenever a boat is under isolation, nobody is permitted to get onto or land. Hence, especially severe limitations are applied by UK specialists to creatures brought into the country.

Quay

A strong design, as a rule of stone or iron close by or projecting into the water, is utilized as an arrival stage and for stacking and dumping ships (Figure 70).

Figure 70. Panoramic style wide-angle view of circular quay from Sydney Harbor Bridge.

Source: Image by Wikimedia Commons.

Quality Assurance Unit

Investigators from the English Tourism Council (ETC) who visit lodgings and give rate marks for quality.

Queen's Awards

Yearly honors to British firms and different associations to perceive exceptional execution in their particular fields, at first for sending out an innovative accomplishment; the Queen's Award for Environmental Achievement was sent off in 1993.

However, following an audit in 1999, the plan was renamed the Queen's Awards for Enterprise. In 2002, 131 associations got grants: 85 for International Trade, 37 for Innovation, 9 for Sustainable Development. In latest years, grants were made to firms in movement, the travel industry and accommodation ventures.

Quicksand

A mass of free fine sand, in some cases blended in with mud, supersaturated with water to be found on certain coasts and close to waterway mouths, which will in general suck down any weighty item, including an individual.

R

Rabies

Intense viral contamination of the sensory system. Side effects incorporate incoherence, loss of motion and difficult muscle fits in the throat; it is generally lethal. It is therefore generally shrunk by being nibbled by a canine or one more contaminated creature and happens in Europe and North America as well as less created nations (LDCs). Hence, the principle of insurance is staying away from contact with creatures and an antibody is likewise accessible.

Rack Rate

Standard full or distributed cost per lodging, to be recognized from different extraordinary (limited) rates at which rooms might be really sold, and furthermore from normal room rate (ARR).

Rail Pass

An exceptional ticket paid for ahead of time, which permits limitless travel via train in a country, throughout a restricted timeframe.

Railway Lost Property Office

An office which gathers objects that individuals have abandoned in trains and keeps them until their proprietors gather them.

Rail (Way) Gauge

The width between the highest point of the rails, which changes in various areas of the planet as follows:

- The standard measure utilized in Europe (with the exception of Spain, Portugal, previous USSR, and certain nations connected to the USSR), North America, and portions of Australia is 1.435 meters (4 ft 81/2 in).
- Broad checks of 1.6 meters (5 ft 3 in.) and 1.65 meters (5 ft 6 in.) are utilized in Spain, Portugal, previous USSR, portions of the Indian Subcontinent, Australia, and South America.
- Narrow checks, particularly 1.066 meters (3 ft 6 in.) or less, are utilized in South Africa and portions of Australia.
- The meter measure (3 ft 33⁄8 in.) is utilized in many areas of the planet (Figure 71).

Figure 71. A narrow-gauge of railway track.

Source: Image by Flickr.

Railway Mania

Portrayal gave in Britain to the 10 years of the 1840s (when a greater number of miles of track were opened than during some other 10 years).

Ramp

A slanting plane joins two degrees of ground. (b) A plane associates a roll-on/roll-off boat to the shore or quay, which might be bow incline (at the front), harsh slope (at the end) or side slope (at the side of the boat). (c) The staircase on wheels was used to stack and dump an airplane.

Ramsar Sites Conservation

Areas of bog and different wetlands of global significance for their untamed life, named after an Iranian town, in which the worldwide show for their assurance was held. Therefore assigned by true nature protection specialists, an aggregate of close to 150 locales existed in the UK by the mid-2000.

Rate of Exchange

The cost at which one money is traded for another. At a specific time without controls, the actual not entirely settled by the organic market for monetary standards in unfamiliar trade markets. Money is therefore said to appreciate/deteriorate while its drifting pace of trade increments/falls concerning different monetary standards. Moreover, changes in fixed paces of trade are called revaluations/debasements. Thus, developments in return rates practice a significant effect on worldwide travel and the travel industry streams.

Readership

The number of individuals who read a paper or periodical is unmistakable from the number who purchase or get it (dissemination). A readership figure can ordinarily be anticipated to be higher than a dissemination figure. Therefore, the two figures are of specific importance to publicists of labor and products, including travel, the travel industry and accommodation items.

Real Terms

Cash esteem at steady costs, i.e., adapted to changes in costs. To dispense with the impacts of cost changes, information at current costs is switched over completely to consistent costs by utilizing record numbers. The travel industry consumption information is therefore ordinarily changed over by utilizing the Retail Price Index (average cost for most everyday items list).

In any case, as these actions change in purchaser costs paid by families, in various nations, a rising use is made of exceptionally developed lists of traveler costs, which reflect all the more precisely vacationer spending designs.

Reality Tourism

To regions of the world considered unstable or less developed, to gain a better understanding of their way of life.

Receiving Country

For purposes of international tourism statistics, the country that receives visitors who are residents of another country.

Recession

Economic activity has declined overall as measured by national income, employment, and other aggregates. A widely accepted technical definition is at least two consecutive quarters of falling output in authentic terms. This has therefore occurred three times in recent years ecumenical: 1974–1975, 1980–1981, 1990–1992, the last being the longest recession since World War II.

Recipe

The creation of a particular dish containing the ingredients and the method of preparation; when utilized in catering. It may also include information concerning the cost of the dish and the nutrients contained therein.

Reconfirmation

Certain airlines require a passenger to confirm the intention to use the reserved seat if the international journey (other than within Europe) is broken for more than a certain time. Failure to reconfirm may result in the abrogation of the next and any subsequent reservation and the seat being sold.

Record Locator

Also called PNR (Passenger Name Record) number, an identification number or code provided by an airline or computer reservation system (CRS) for each booking.

Recovery Rate

A performance metric used by public and nonprofit organizations, such as tourism/visitor magnetization, in which total income is divided by total operating expenditures (3,100).

Recreation

Particular utilization of leisure or activity was undertaken during leisure, which may include travel and tourism. Some major distinctions are indoor/ alfresco, home-predicated/away from home, and active/passive recreation.

Recreation(al) Vehicle (RV)

A motorized caravan or motor home for holiday use, as well as a truck, van, or any other type of conveyance that can be used off-the-road, such as a dune buggy, or an off-road motor bicycle, or a snowmobile. The latter three are often a source of conflict with conservationists because of the damage and disturbance they cause, to citizens, and to those who seek mute delectation in remote areas.

Referral

In general, the act of one person or organization recommending another, with or without payment, is known as a referral fee or referral commission. For example, each hotel in a group may promote other hotels in the group and engender business for them by onward reservations. Hence a hotel consortium is sometimes referred to as a referral system, especially in the USA.

Refugee

The United Nations, in the 1951 Convention relating to the Status of Refugees, elongated in its application by the 1967 Protocol relating to the Status of Refugees, defined a refugee as a person who 'owing to the well-founded trepidation of being persecuted for the reason of race, religion, nationality, membership of a particular gregarious group or political opinion, is outside the country of his [or her] nationality and unable or, owing to such fear, is indisposed to avail himself [or herself] of the aegis of the country.'

Regatta

This is a race on a sailing vessel or yacht, or commonly a series of such races, which is a sporting and social event deriving from certain races on the Grand Canal in Venice. Eminent regattas, such as the Henley Regatta in England, are major tourist/visitor magnetizations.

Region

A major area within a country, which has certain attributes in prevalence, such as climate or topography, and then is customarily described as a natural region, and/or which forms a unit for political or administrative purposes. Catalonia in Spain, the Highlands in Scotland, Languedoc-Roussillon in France, are well-kenned examples in Europe. (b) An area of the world with defined characteristics or a group of countries in geographical proximity, e.g., the Balkans, the Caribbean, Middle East (ME).

Regional Development

It consists of the major economic and societal development of a defined area, traditionally an administrative entity, particularly one suffering problems, in which the regime, directly or indirectly, plays a significant role by stimulating, orchestrating, coordinating, and/or financially boosting that development. Moreover, in areas with climatic, scenic, or other magnetizations and deficient in resources for other forms of economic activity, tourism often represents a paramount element in regional development, as is the case in many coastal and mountainous regions.

Regional Planning

A systematic and comprehensive approach to orchestrating the economic and convivial development of a defined major area, customarily undertaken by regimes, for areas that represent administrative units and are broader in scope than statutory land use orchestrating areas. Some of the best-known examples of regional orchestrating of tourism paramountcy in Europe have been on the Adriatic coast of former Yugoslavia, Languedoc-Roussillon in France, and components of the Iberian coast.

Regional Tourism Authorities (RTA)

Seven regional organizations in the Republic of Ireland with a membership of local ascendant entities and individuals, sodalities, and firms, providing visitor servicing, regional marketing, and development coordination and orchestrating.

Regional Tourism Organization

Intermediate level of tourism organization between national and local levels, variously concerned with the development, promotion, and coordination of tourism in its area. Therefore, usually perceive, e.g., Regional Tourist Boards (RTBs) for England, area tourism companies (ATCs) for Wales, Regional Tourist associations (RTAs) for Northern Ireland, Regional Tourist Authorities (RTAs) for the Republic of Ireland.

Regional Tourist Associations (RTAS)

An organization whose members come from the public and private sectors and operates under the auspices of the Northern Ireland Tourist Board (NITB) intending to develop, promote, and coordinate tourism in their respective areas of Northern Ireland.

Regional Tourist Boards (RTBS)

Network of ten voluntary bodies with a tripartite structure of local ascendant entities, the tourism industry, and the English Tourism Council (ETC) covering the whole of England. Moreover, composed following the establishment of the national board for England under the Development of Tourism Act 1969, to develop, promote, and coordinate tourism in their respective areas of England.

Regression Analysis

A statistical technique for establishing a relationship between a dependent variable and one or more independent variables, which would expound on past variations in the former and prognosticate future variations in terms of changes in the latter. A simple regression model is utilized for a two-variable relationship, such as holiday (vacation) participation and income. Further, multivariate regression models are required for more than two variables. E.g., to forecast tourist flows to several destinations, population size, and income per capita of each of the engendering countries, relative distance and peregrinate times may be used.

Religious Travel/Tourism

In a narrow sense, trips, and visits whose main purpose is the religious experience, e.g., pilgrimages to Jerusalem, Lourdes, and Mecca. In a broad sense, withal trips and visits whose major motivation is religious heritage, such as churches and cathedrals.

Remittances

The capital sent by immigrants from the country in which they work to relatives in their country of inception; these amounts appear as 'private transfers' among Invisibles in the balance of payments current account.

Repeat Customer

Customer who buys products or accommodations from the same supplier, e.g., guest staying at the same hotel or peregrinator utilizing the same airline. Hence reiterate tourist or visitor returns to the same destination.

Resale Price Maintenance (Rpm)

Suppliers set categorical or minimum prices for their products and require distributors to sell them at those prices. The practice is subject to legislation in many countries. E.g., in the UK all resale price arrangements are surmised to be against the public interest unless proved otherwise, and tour operators cannot licitly enforce inclusive tour (IT) prices against retail travel agents. However, suppliers may publish a recommended retail price (RRP) and concur with distributors to what extent such prices may be discounted.

Reservation Diary

Each day is represented by a page in a ledger, showing a list of all the rooms in a hotel and marking if they are booked or vacant.

Reservoir

An artificial body of water, engendered in upland valleys by the construction of a dam or barrage, and in the lowlands by the building of a wall or bund to enclose the water. Reservoirs are engendered for several different purposes and these and their location and characteristics affect their suitability for recreational use: direct supply of imbibing water, control of flow in a river for extraction lower down, generation of hydroelectric puissance, irrigation, and the maintenance of water levels in canals. Moreover, apart from the first, there is no reason why the water should not be utilized for recreation, and modern purification plant makes some forms of recreation acceptable on the direct supply reservoirs, eminently sailing and angling (Figure 72).

Figure 72. Aerial photo of the Saugatuck Reservoir.

Source: Image by Wikimedia Commons.

Resident

For purposes of international tourism statistics, a person is considered to be a citizen in a country if the person has lived for most of the past year (12 months) in that country, or (b) has lived in that country for a shorter period and intends to return within 12 months to live in that country.

For purposes of statistics of domestic tourism, a person is considered to be a citizen in a place if the person: (I) has lived for most of the past year (12 months) in that place; or (II) has lived in that place for a shorter period and intends to return within 12 months to live in that place [World Tourism Organization (WTO)].

Residential Hotel

An accommodation that accommodates long-term guests who may choose to make it their home.

Resort

A place to which people go for holidays (vacations) and recreation, hence holiday (vacation) and health resorts, additionally inland and coastal/seaside resorts. Historically the evolution of tourism has been proximately identified with the commencements and subsequent development of resorts.

Nowadays the term often has its literal meaning to denote any visitor center to which people resort in astronomically immense numbers and capital cities incline to be the most astronomically immense and most prosperous resorts in their countries, especially for international tourists. Hence, I n the USA and the Caribbean, additionally, a holiday (vacation) hotel provides extensive regalement and recreation facilities.

Resort Representative

A tour operator's employee is predicated in a resort and providing a formal point of contact between the firm and clients and resort facilities and accommodations, as well as regalement and other arrangements.

Resort Tax

In the form of bed, hotel, or room tax, it is a tax imposed by local authorities or other agencies on visitors to raise revenue. Sometimes the proceeds go to tourism.

Resource-Based Resources

Resources devoted to recreation and tourism, depend for their magnetization on their quality irrespective of their location. Their character magnetizes from a considerable distance and their appeal is national or international rather than local or regional.

Restaurant Establishment

Providing food for consumption on the premises to the general public, to which the supply of alcoholic liquor, if any, is ancillary, as a separate unit, or as a component of a hotel or another establishment. Beyond this generalization, some restaurants operate under designations such as cafés, snack bars, and the like. Moreover, in some countries, the designation of restaurants and other consuming establishments is regulated by law. In 2000 there were around 45,000 restaurants, cafés, and takeaway food shops operating in the UK.

Rest Area

Parking areas near motorways or other astronomically large roads, with public restrooms, picnic tables, and benches, where you can park and unwind after a long drive.

Retail Export Scheme

Scheme administered by Customs and Excise in the UK, which enables overseas visitors to receive restitution of value integrated tax (VAT) paid on some goods bought while in the UK. From 1st January 1993, European Community (EC) travelers have no longer be eligible to utilize the scheme, but it perpetuates to be available to others and additionally in certain circumstances to UK residents departing abroad and to crew members of ships and aircraft.

Revalidation Sticker

An amendment is attached to the flight coupon of an airline ticket, exhibiting a vicissitude such as a change of flight mode to the pristine reservation.

Ribbon Development

Building along main roads elongating outwards from built-up areas, a common direction of much urban magnification, including many coastal and inland resorts. Additionally used to describe more generally a linear building development along a coastline, valley, or route.

Rickshaw

Traditionally used as a form of tourist transportation in Hong Kong and other Asian cities, this light two-wheeled carriage is drawn by people standing between two shafts.

Right of Way

A licit right of passage across the property of another, conventionally by a defined footpath. Of particular consequentiality in Great Britain (GB) where there is a dense network of footpaths throughout areas utilized for farming, representing a major recreational resource.

Ring-and-Ride

A term utilized in community conveys a scheme that enables users such as incapacitated and elderly people to telephone and arranges for door-to-door convey, conventionally by taxi or minibus.

Riviera

The Mediterranean coastal region faces the Ligurian Sea and elongates between the departments of Alpes-Maritimes in southern France and Liguria in northern Italy. Therefore, the French Riviera is known as the Côte d'Azur and includes such well-kenned resorts as Nice, Cannes, and Monte Carlo in the Principality of Monaco. The term Riviera is sometimes withal applied to other paramount tourism areas with a captivating climate and scenery, conventionally for marketing purposes, e.g., Cornish Riviera in England.

Room Safe

Instead of the centralized safe custody facility provided through the hotel reception, hotel bedrooms may also be equipped with a guest-operated safe, where guests might keep their valuables.

Room Tax

Tax levied by the central or local regime or another agency on staying visitors, accumulated at the place of stay, as an expedient of raising revenue; sometimes the proceeds are applied to tourism purposes. May be additionally called bed tax, hotel tax, or visitor tax.

Rooming List

List of names of members of a group provided by the travel organizer in advance of advent and utilized by the hotel to assign rooms.

Round-the-World Ticket

Long haul airline tickets are predicated on the cumulated networks of two or more airlines. Customarily available for economy, business, and first-class travel and valid for a year, sanctioning travel in one direction with stopovers.

Round Trip

A synonym for a return journey. For civil aviation purposes, the International Air Convey Sodality (IATA) defines round trip as travel from a point of inchoation to another turnaround point and return via the same air route used outbound, regardless of whether the outbound and inbound fares are identical, or by an air route different from that used outbound, for which the same common all year one-way fare subsists.

Route

A way, road, course; a certain direction taken in traveling from one place to another. (b) To travel, send or forward by a certain route as, e.g., (in aviation parlance) AT (via the Atlantic) or AF (via Africa).

Rural Tourism

For the most part occasion (excursion) the travel industry and emphasis on wide-open objections. Albeit not another peculiarity, country the travel industry advancement has been getting a lot of consideration as of late for of pay and work age and of broadening of nearby economies. It is therefore frequently related to cultivating the travel industry yet provincial the travel industry is a more extensive idea and incorporates additionally such unambiguous elements as the improvement of nature trails, cookout destinations, translation focuses, and rural and society exhibition halls (Figure 73).

Figure 73. Xiaodubai rural tourism area.

Source: Image by Wikimedia Commons.

Russian Service

Style of café table assistance, in which the food is divided and put on silver salvers in the kitchen, and served in the eatery from the salvers onto the visitor's plate.

Ryokan

A conventional Japanese hotel, usually with a nursery, private suites, or visitor room with shared restrooms, and suppers regularly served in rooms.

S

Safari

Originally denoting a hunting excursion in Africa, safaris are now additionally outings to observe untamed life, especially in East and South Africa and are sometimes referred to as photography safaris.

Safe Custody

The office is presented by banks to their clients and by inns to their remaining visitors to store resources and reports for safety's sake.

Sailing Card/List

For every one of its vessels, a transportation organization will give a schedule containing the dates and locations of its sailings, cruises, and appearances at ports at which the ship will stop.

Sale and Lease-Back

A game plan by which an organization offers its territory and structures to a financial backer and leases a similar property back for a concurred term. Moreover, a typical method for supporting an inn venture, the impact is to deliver capital restricted in the property for different purposes, as the lodging organization is then ordinarily simply expected to fund interest in inside resources and to pay a lease.

Sales Promotion

Generally, activities are designed to stimulate sales other than advertising and publicity. Therefore, common incentive sales promotions in travel, tourism, and hospitality include price-cutting, discount vouchers, extras, gifts, and frequent user programs, as well as point-of-sale promotions.

Salvage

The sale of waste or outdated materials, such as used cooking oil, squandered paper, and other material, in hotels and restaurants is an important source of income.

Same-Day Visitor

For factual purposes, a guest (either homegrown or global), or traveler, who does not go through the night in aggregate or private convenience in the spot visited [World Tourism Organization (WTO)].

Sampling

A technique for request where information is gathered from an extent of the populace to give data all in all populace, as unmistakable from a registration, wherein the information is gathered from the entire populace.

Whenever every individual or thing has an equivalent possibility being picked, the example is known as a straightforward irregular example; when the populace comprises of different gatherings, a separated arbitrary example might be utilized, i.e., determined extents of the example are drawn indiscriminately from various gatherings or layers. Hence, test reviews are the primary type of overview in movement and the travel industry.

Sanatorium Establishment

For the treatment of invalids, particularly consumptives and convalescents, regularly situated in high elevations and different areas with health-giving properties, e.g., the Alps and along the ocean coasts.

Satellite Account

Term created by the United Nations to quantify the size and construction of monetary areas excluded from their right in public records. Consequently, a travel industry satellite record estimates the travel industry area. Further, the record runs close by as a subset of public records and empowers the arrangement of solid assessments of the commitment of the travel industry to the nation's economy, which can be precisely contrasted and different areas.

Scan Rail Pass

Limitless travel rail tickets are accessible for different periods in first or standard class to non-occupants of Scandinavia on the public railroads of Denmark, Finland, Norway, and Sweden and certain different administrations. Various costs apply to those under 12, 12 25, 26–60, and 60+ years old.

Scenic Route

For the most part, a minor street going through an alluring field is moderately minimal utilized by business traffic and thus appropriate to joy motoring. Frequently recognized by nearby specialists and advertised as a traveler/guest fascination yet additionally to draw in vacationers off significant streets.

Schengen Agreement

Deal endorsed in Luxembourg town of that name in June 1985 by France, Germany, and the three Benelux nations on the progressive cancelation of controls at the normal boondocks. It required over 10 years for it to come into force. Moreover, the Schengen zone covers 13 EU states, including Denmark, Sweden, and Finland, and two non-EU nations (Norway and Iceland), which participated in March 2001, as a free travel area of 310 million individuals permitting crossing borders without showing international IDs. The UK and the Republic of Ireland have quit the greater part of the understanding. The framework additionally incorporates a data set to battle wrongdoing and unlawful movement.

Seamen's Fares

Decreased air admissions to which sailors are generally qualified while going to join or leave a ship.

Season

A division of the year is ordinarily connected with the environment and related factors. There are four times of comparative length in the center scopes, yet two in the tropics (wet and dry) and the polar areas (winter and summer), and three in the monsoon regions (cool, sweltering, and wet); there is little separation of the season in central districts.

Seasonal Adjustment

The cycle of changing factual time series to eliminate variances that show a normal occasional example throughout the year, and which might be assessed from earlier years. The changed series gives a clearer perspective on the basic pattern, albeit any excess changes may likewise contain repetitive developments, as well as unpredictable and outstanding varieties. Therefore, like many travelers, the travel industry and accommodation administrations are dependent upon irregularity. Hence, occasional change is generally utilized in related month to month and quarterly measurements.

Seasonality

Variety is indicated by the time or seasons of the year, which will in general display a comparative example from one year to another, and influence exercises like agriculture, development, the travel industry, and related business. Moreover, when related to climatic and related factors, this is now and again portrayed as normal irregularity, to separate it from institutional irregularity. For example, that reflected in the schedule of the chapels and such occasions as journeys. However, a significant part of institutional irregularity is examples of occasion (vacation) taking, to some extent directed by school occasions and impressively reducing the period when climatic circumstances are favorable.

Seat Sale

Unique return air charge presented via aircraft on European planned trips for travel during times of low interest, dependent upon different limitations, for a maximum stay of one month. Cannot be changed or dropped.

Seat Turnover

A proportion of usage of a catering establishment limit shows the times each seat is utilized by clients throughout some undefined time frame. For example, when in a café seating 100 clients, 250 are served in a day, the day-to-day seat turnover is 2.5.

Second-Home

A house, loft, or another building is utilized for occasions (get-aways) and other transitory stays by proprietors whose fundamental home is ordinarily somewhere else. Otherwise called holiday/summer/getaway home. Therefore, second homes are likewise much of the time utilized by proprietors' friends and family members and let by proprietors to other people. Hence, second house buying is exceptionally evolved in North America, a few pieces of Eastern and Western Europe, including France and Scandinavia, and, on the off chance that static parades are incorporated, likewise in Great Britain (GB).

Sector

Part of the economy, e.g., private area, public area, the travel industry area, by and large, contains various enterprises. (b) A piece of an excursion via air, which might comprise at least one leg or a section. A leg is a piece of an excursion between two sequential booked stops on a specific flight. Therefore, a section is the piece of an excursion on a specific departure from the traveler's boarding point to the disembarkation point; a fragment might comprise at least one leg.

Security Tax

The tax collected by numerous legislatures and air terminal specialists at air terminals to assist with settling extra security costs lately. The term is likewise at times utilized for a security overcharge imposed via aircraft to make up for expanded safety efforts as, e.g., by certain US carriers on transatlantic flights.

Self-Catering Accommodation

Short-term accommodation is provided without meals in many holiday habitats, leased rooms, houses, lofts, and comparative foundations, where cooking and different facilities are accommodated for visitors' utilization. Therefore, huge self-catering edifices frequently give retail outlets to arrangements, as well as cafés as discretionary offices.

Self-Employed

One employed in his/her principal work on his/her record, regardless of any representatives. An enormous piece of work in movement, the travel industry, and hospitality is in private companies and incorporates numerous independently employed; even though there has been a development in the centralization of possession and in the size of activity. Hence, independent companies and the independently employed keep on being conspicuous.

Semiotics

Otherwise called semiology, the investigation of signs or how items, words, or pictures convey ideas or pictures. Of specific importance in publicizing, semiotics commits to make in the advancement of tourist items and the travel industry locations (Figure 74).

Figure 74. Semiotics of the kitchen.

Source: Image by Wikimedia Commons.

Service Charge

A rate (for the most part somewhere in the range of 10% and 15%) is put on hotels and eatery tabs as a tip in place of direct tipping. Moreover, service compris, a French term, indicates that the gratuity is included with the price of the room or meal.

Serviced Accommodation

Short-term convenience is typically furnished by inns and similar foundations with dinners and administration, and frequently by holiday centers and others.

Seventh Freedom

In civil aviation a right (not designated in the Chicago Convention) to operate stand-alone accommodations entirely outside the territory of the airline's home country, to carry traffic between two foreign countries. A US airline had such rights for many years to operate a shuttle accommodation between Tokyo and Seoul. Therefore, the incidence is incrementing in Europe under European Amalgamation liberalization measures. Moreover, one of the first Seventh liberation rights has been taken up by the Belgian airline Sabena to fly between Barcelona and Venice.

Sex Tourism

Outings and visits are fundamentally propelled by the possibility of sexual experiences, a huge movement in various emerging nations, particularly in South-East Asia.

Shared Commission

Commission got from administrators and shared by non-selected travel planners who acquire tickets from delegated specialists. Additionally called split commission.

Shore Excursion

A visit accommodated journey travelers at a planned stop on a voyage, for the most part by transport or mentor, typically discretionary and therefore paid for independently.

Short Holiday (Vacation)

A term differently characterized for specific purposes, e.g., in most public holidays (get-away)/travel/the travel industry overviews in Europe as occasion (get-away) of one to three evenings/two to four days from home. Likewise, in some cases depicted as brief breaks. Therefore, lately, short occasions addressed over a portion of all occasions of UK inhabitants in the UK [United Kingdom Tourism Survey].

Shoulder Periods

Seasons between peak and low seasons are frequently distinguished as such by suppliers of vacationer/guest attractions, offices, and administrations and reflected in costs, which are lower than in the peak and higher than in the low season.

Shuttle

Successive vehicle administration between the two focuses on high-density routes for which no reservations are required. Therefore, aircrafts working transport benefits ordinarily offer late registration times near flight takeoff, and some likewise back-up assistance on the off chance that a specific flight is full as, e.g., British Airways among London and a few UK urban areas.

Sightseeing

One of the most well-known types of passive recreation in developed nations and a significant holiday (get-away) movement by walking, via vehicle, in mentor visits, travels, journeys from occasion bases, and travel in little gatherings or alone utilizing an assortment of transport. Moreover, the attractions might be alluring open country, fabulous views, archeological destinations and authentic landmarks, natural life, and the lifestyle of others. Vacationers are progressively brave in how far they travel to see uncommon elements, like Antarctica, Galapagos Islands, or the Silk Road.

Silk Road

Name given to various memorable shipping lanes from East China to focal Asia however presently a tourist course via train and transport from Xian, through Langzhou and Urumchi, to Almaty (Alma Ata) or Tashkent (Figure 75).

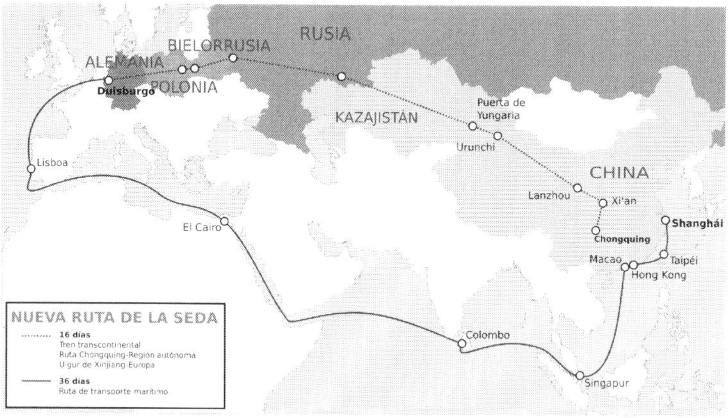

Figure 75. *Silk road.*

Source: Image by Wikimedia Commons.

Single Supplement

Extra installment by a visitor for single inhabitance of a lodging, typically on a coordinated visit for which charges are indicated based on sharing a room.

Sleeper Seat/Sleeperette

A reclining transport seat has typically been given in the first-class cabin on long stretch flights, intended to be flexible to a sleeping position. Also called dormette.

Snowline

The most reduced degree of super durable snow cover in mountain regions, changes with the area, orientation, environment, and geography. Nonetheless, on the northern side of the equator, a large part of the lower ground is additionally snow-canvassed in winter, giving open doors to an assortment of snow-based amusements.

Snowmobiling

Movement over snow, essentially in the marshes, on a mechanized sledge, equipped for going at speed additionally utilized as a broader type of transport in winter and in for all time snow-covered regions. Moreover, normal for both the travel industry regions and the more settled regions, and frequently a wellspring of contention with different types of amusement and with occupants due to clamor contamination and antagonistic consequences for different types of traffic. Notwithstanding the utilization of ways and tracks, travel by snowmobile happens over snow-shrouded ranch land, where mishaps have happened due to inconspicuous walls.

Social Tourism

The travel industry interest of individuals of restricted means and those distraught through age, incapacity, or family conditions, empowered, and made conceivable by exceptional measures. These remember for training sponsorships, specific offices, and different measures, at times of a helpful sort, once in a while by the state or another outsider. Moreover, social tourism is profoundly evolved in Belgium, France, and Eastern European nations specifically.

Soft Class

Term utilized in certain pieces of Asia for agreeable more costly seating on trains, rather than a hard class to allude to the least expensive and most fundamental classification.

Spa

A retreat with mineral or warm water utilized for drinking and washing; in American speech, any hotel giving medical care offices and administrations. The term gets from the name of a town in Belgium and has been applied to resorts around the world, as well as being integrated with a considerable lot of their names, particularly in Europe. Therefore, the restoration and development in the fame of spas from the finish of the sixteenth century onwards, and their change into popular delight resorts. Hence, it addresses along with the Grand Tour and the rise and development of coastal resorts, the beginnings of the travel industry as far as we might be concerned today.

Space Travel/Tourism

Excursions and visits to space, the first having been embraced by a paying American in a Russian container to a space station in 2001.

Special Group Inclusive Tour Fare (SGIT)

European comprehensive tour (IT) fare, in the UK accessible to visit administrators holding the air travel organizer's license (ATOL), which might be utilized for gatherings of at the very least ten travelers.

Special Interest Holidays (Vacations)

A conventional term for holidays (excursions) for individuals with specific interests, like expressions, schooling, different leisure activities, sports, and outside exercises. These are a portion of the fastest-growing fragments of the travel industry, with numerous expert suppliers.

Split Shift

A term utilized in inns and eateries, yet additionally in other businesses, for a working example of two working periods isolated by a long stretch between them (i.e., longer than an ordinary rest or dinner time), e.g., 11 a.m. to 3 p.m. what's more, 6–10 p.m.

Sports Tourism

Trips and visits are motivated by participation in sports or by spectator attendance at sporting events.

Stabilizer

A restrictive managing plan was presented in 1965 by the Association of British Travel Agents (ABTA) by which visit administrators would sell their visits just through ABTA travel planners and ABTA travel planners wouldn't sell voyages through non-ABTA visit administrators. Moreover, following a test by the Office of Fair Trading, Stabilizer was maintained by the Restrictive Practices Court in 1982 as working in the public interest. Notwithstanding, by 1993 the requirement for its continuation turned out to be progressively far-fetched because of the EC Package Travel Directive, the execution of which made the security managed by Stabilizer unnecessary, and its destruction followed.

Standard International Classification of Tourism Activities (SICTA)

A supply-based construction of the travel industry was created by the World Tourism Organization (WTO) and embraced as a temporary grouping by the United Nations Statistical Commission in March 1993. SICTA was distributed together by the UN Statistical Commission and the WTO in the full report on Recommendations on Tourism Statistics in 1994.

Standby

In travel and the tourism industry, most generally utilized comparable to an eventual traveler without a booking taking a risk and trusting that a seat will open up; subsequently backup traveler, likewise reserve admission, a diminished charge accessible for reserve travelers on certain courses.

Stopover

For the most part, a planned break in an excursion concurred ahead of time by the transporter. What establishes a visit for specific purposes might be authoritatively characterized, as, e.g., in air transport by the International Air Transport Association (IATA). Likewise called layover.

Strath

Scottish expression for a wide valley or low-lying level land limited by slopes or strategic position, for the most part, crossed by a stream and developed. For example, Strathspey, the spot name for the strath of the waterway Spey (Figure 76).

Figure 76. Strath Ossian.

Source: Image by Flickr.

Suite

Set of interfacing rooms comprising of at least one or more bedrooms, washrooms, and lounges and at times likewise extra rooms, for example, a dining area. Additionally used to depict an exceptionally enormous room with isolated living and resting regions called a junior suite.

Supplementary Tourist Accommodation

The term utilized by the travel industry associations ordinarily to incorporate youth lodgings, holiday centers and towns, leased rooms, houses, and flats, and camping areas. Nonetheless, the specific importance will in general contrast among nations, and mirrors the assignments given to convenience foundations in every country.

Surcharge

Beneficial charge added to the air admission by the aircraft, comprehensive visit cost by the visit administrator, or another travel buy by the administrator, as a rule, to make up for expansions in costs between the hour of booking and the hour of movement. However, for instance, a cash overcharge is exacted to make up for vacillations in return rates, a fuel overcharge for expansions at the expense of flight fuel, and a security overcharge for inflated cost of safety efforts.

Survey of Visits to Tourist Attractions

Annual astronomically immense-scale sample survey carried out by the Department for Culture, Media, and Sport, sponsored by British Tourist Ascendancy and National Tourist Boards, on several visits, admission charges, revenue, employment, and expenditure since 1976. Results are published in Digest of Tourist Statistics and Sightseeing in the UK.

Sustainable Tourism

Emerging from concepts of sustainable development, as defined in the 1987 report of the United Nations World Commission on Environment and Development, Our Prevalent Future (the Brundtland Report), the term may be applied to all forms of tourism that are in harmony with their physical, gregarious, and cultural environments in the long term. Moreover, not confined to minuscule-scale, as alternative tourism, sustainable tourism development may be viewed in terms of the Brundtland Report as development that meets the desiderata of the present without compromising the activity of the future generations to meet their own needs.

Switch Selling

The act of drawing in clients by dealing with low-priced, non-existent, or non-accessible labor and products, in order to achieve an offer of another option. Called bait and switch in America, the training is prohibited by regulation as well as a governing set of principles in numerous nations.

SWOT (Strengths, Weakness, Opportunities, and Threats)

Strengths, weakness, opportunities, and threats (SWOT) is utilized as a framework for analyzing information about an organization as a component of developing its marketing strategy and tactics. Strengths and impuissance's include respectively present and future advantages and disadvantages of the organization vis-à-vis its competitors. Furthermore, opportunities and threats are those identifiable in the present and future operating environment of the organization. Thus, the same framework may be utilized for analyzing a project, site, or area as a component of an appraisal of its tourism potential.

Syndicated Survey

A survey commissioned by or on behalf of a group of clients on a cost-sharing substratum, in which all or some of the results are supplied to different clients, and in which individual clients may commonly include their questions. This is therefore significantly less extravagant for the participants than each carrying out their survey, and a cost efficacious approach, especially for more minuscule businesses. British National travel Survey (BNTS) is a major example of a syndicated survey in tourism.

Synergy

Term portraying cooperating/coordinating when the consolidated activity makes up an entire more prominent than the amount of the different parties.

System One

US computer reservation systems (CRS) are claimed by Texas Air (Eastern and Continental Airlines) as a joint endeavor with General Motors' Electronic Data Services.

T

Table Tent

Tent cards are printed and folded cards that are hung on restaurant tables, bar counters, and front counters of hotels for market-specific food and drink items and house services.

Table Wine

Wine aged to approx. 12–14% liquor and considered appropriate for drinking with suppers.

Tachograph

An instrument for recording the distance voyaged and the speed of transport, mentor, or truck.

Tacis

A project to provide technical assistance to the commonwealth of independent states (CIS) to help them build stable market economies, is part of the European Cumulation Project.

Tactics

Techniques and approaches utilized in executing a system, ordinarily moderately short term. Thus, strategic promotion, which includes dealing with the showcasing blend.

Tailwind

An increase in speed is caused by wind blowing from behind a ship or aircraft.

Tapas

Spanish expression for little plates of tidbits, like fish, cheddar, and olives, normally presented with beer or wine. Consequently, tapas bar which works in serving tapas (Figure 77).

Figure 77. Tapas En Barcelona.

Source: Image by Wikimedia Commons.

Target Marketing

Focusing the marketing effort on a pellucidly defined group or groups of potential customers chosen for categorical marketing attention (target market), which is generally identified by segmentation. Additionally known as niche marketing. Target markets in peregrinate, tourism, and hospitality are commonly defined by reference to demographic, socio-economic, and psychographic criteria.

Tarn

A miniature lake surrounded by mountains, fed by rainwater from the steep slopes it circumvents instead of a separate tributary.

Teleconferencing

Meeting in which three or more participants in different locations communicate with each other through telecommunications. They perceive each other aurally in audio conferencing, as with telephone communication; but in videoconferencing. Therefore, they perceive each other both visually and aurally through a satellite or other network. The system thus enables people to hold meetings without the time and expense of traveling.

Telemarketing/Telephone Marketing

Any showcasing action in which the phone is utilized to contact individuals and to look for data (phone promoting research) or to request orders (telephone selling).

Teletex

A data transmission framework however quicker than the message, which permits, i.e., a direct association between word processors.

Teleworking

Individuals telecommuting utilizing data innovation, a developing practice in created nations. E.g., during the 1990s, more than 10% of managers in Great Britain (GB) utilized specialists based at home and more were intending to do as such.

Telex

Automated transmission system for sending printed messages nationally and internationally between subscribers to the system. The rental conventionally covers equipment and maintenance accommodation; telex use is charged to subscribers on a tariff substructure kindred to that for telephones.

Temperate Zones

Broad latitudinal temperature belts between the Arctic Circle and the Tropic of Cancer (Northern Temperate Zone) and between the Antarctic Circle and the Tropic of Capricorn (South Temperate Zone). Moreover, the North Zone includes the bulk of Europe and all continental USA except Alaska, the South Zone much of Australia, the whole of Incipient Zealand, and the bulk of South Africa.

Temporary Visitor

Any person, without distinction as to race, sex, language, or religion, who disembarks and enters the territory of an ICAO contracting state other than that in which that person customarily resides, and remains there for not more than three months for legitimate non-immigrant purposes [International Civil Aviation Organization (ICAO)].

Tempus III

The third phase of the European Amalgamation program is designed to fortify curricular development and structural reform in higher inculcation. The program is therefore open to the non-associated countries of central and eastern Europe and 13 TACIS countries (12 NIS and Mongolia) in sodality with the 15 EU member states.

Tender

- The vessel that transports passengers and crew between a ship and shore when it is anchored offshore, as well as stores;
- The formal offer, usually in response to a request for tenders, to perform work or deliver goods and services at a stated price;
- Currency, as in legal tender, i.e., the currency that cannot be legally refused in payment of a debt.

Terminal

- Traveler and additionally freight region toward the finish of transport courses where travelers set out and land and freight are stacked and dumped. Air transport terminals in downtown areas are known as air terminals or city terminals.

- Computer equipment normally consists of a keyboard and visual display unit (VDU) screen linked to the computer, which enables the operator to input data and instructions and obtains output.

Terms of Trade

- Agreements under which business is directed, normally set out by the dealer, and including the privileges and commitments of the purchaser. Such terms cover, e.g., techniques for installment, limits, discounts, and scratch-off expenses, as well as other separate privileges and liabilities of the gatherings.
- In economics, the ratio of the index of export prices to the index of import prices is used as a measure of the trading prospects of a country in international trade. Moreover, export prices tend to rise faster or fall slower than import prices when terms of trade improve, and vice versa.

Territorial Waters

Waters are under the jurisdiction of a state, conventionally within three miles of the seashore, albeit some countries have claimed longer distances. Moreover, Inter alia, fishing in such waters is reserved for the benefit of the subjects of the adjoining state.

Tetanus

A disease of the muscles and nervous system, causing spasms of the neck and jaw; can be fatal. This disease is caused by the invasion of bacterial spores present in the soil, manure, and dust through a wound. Therefore, the peril is ecumenical. Particularly when traveling to rural areas, vaccinations are the only precaution to take (Figure 78).

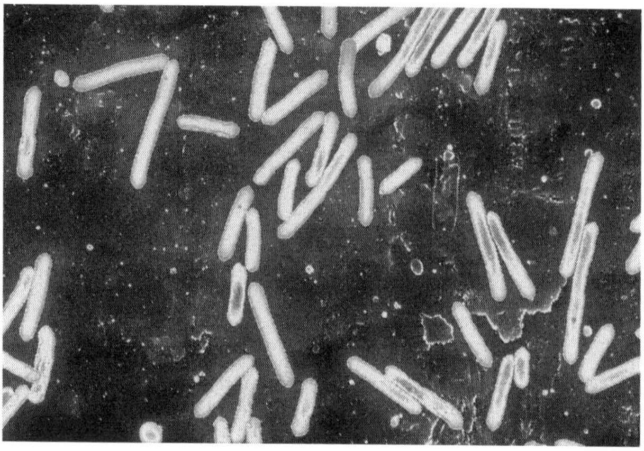

Figure 78. Microscope image of Tetanus bacilli (Clostrium tetani) that cause tetanus.

Source: Image by Flickr.

Thalassotherapy

The treatment is presented in various centers in the Mediterranean and different hotels in southern Europe, utilizing seawater, ocean mud, and kelp to renew and fortify the body.

Thanatourism

The practice of going to locations where death is associated, such as sites of murder and atrocities, and battlefields, is also known as 'dark tourism.'

Thanksgiving

American national holiday celebrating the first harvest of settlers, on the fourth Thursday in November.

Theme Park

A site assigned and worked to give diversion and amusement to guests through an assortment of attractions in light of at least one authentic or different topic, including such administrations as catering and shopping, and for the most part, charging for confirmation. Furthermore, notable models are Disney Parks in the USA, Japan, and France, Europa Park in Germany, and Alton Towers in the UK.

Third Age

Term of generally late beginning to portray the populace bunch already normally restricted to the resigned. Frequently considered to be one of three periods of the human existence range, the first (under 25) being related to learning, the second (25–50 years) with extraordinary business, the third (north of 50) with steady withdrawal from work. Hence, third age is of specific significance in movement, the travel industry, and cordiality in light of the size and development of the gathering, as well as its developing action, wealth, and accessible time in which to partake.

Third World

Alfred Savry, a French economist, coined the term 'non-aligned countries' in the 1950s to describe those committed neither to the market economy block of the West (First World) nor to the state-run economy bloc of Eastern Europe and the Soviet Union (Second World). Nowadays a synonym for developing countries.

Threshold

As an economic concept, a term commonly applied to the lowest demand indispensable to fortify an activity. Quantified in terms of population or income, it is utilized, e.g., in culling a location for a man-made visitor magnetization or a retail outlet.

Throwaway

An item in an inclusive tour (IT), which is infrequently utilized by the customer, is included mainly to qualify for a tour-predicating fare. For example, a person visiting friends or relatives and staying with them may find it less sumptuous to buy an IT and discard the accommodation element than paying any other fare.

Tick-Borne Encephalitis

A viral infection causes an inflammation of the brain, contracted from the bite of an infected tick. Occurs in forested parts of central, eastern, and northern Europe, including Scandinavia, especially in late spring and early summer. Moreover, walkers and campers in the affected areas are advised to wear clothing that covers most of the skin, use insect repellents, and a vaccine is also recommended.

Ticket Point Mileage (TPM)

The total number of miles flown in an airline journey is computed as the sum of the miles flown between each point en-route.

Ticket Tout

An individual who offers admission to popular events at a higher price than the published one.

Ticketing Codes

One-letter standard abbreviations of the International Air Transport Association (IATA) to show various characteristics of a flight on an airline ticket, such as class of service (e.g., F for first, C for business/club, Y for economy); timing (e.g., N for night, W for weekend, X for weekday); types of fares (e.g., E for excursion, P for family, Z for youth fares).

Ticketing Time Limits

Minimum time limits for ticketing concurred by the International Air Convey Sodality (IATA):

- For reservations made more than two months afore peregrinate, at least 30 days afore departure date;
- For reservations made between two months and 15 days afore peregrinate, at least seven days afore departure;
- For reservations made between 14 and three days afore peregrinate, at least 48 hours before departure;
- For reservations made within three days of peregrinate, as anon as possible.

Tidal Range

The difference in ascension between high and low watermarks in coastal areas is affected by orientation and the nature of the coast and the sea bed. Thus, the Mediterranean Sea is virtually tideless, presenting quandaries for keeping beaches clean but auxiliary for navigation, launching craft, and utilizing beaches at all times. However, in contrast, the funnel-like Bay of Fundy in eastern Canada has a tidal range of up to 50 feet (16 m) at its head.

Tied House

In Great Britain (GB), the term is most commonly applied to a public house owned by a brewery and leased to a tenant who is operating the business personally but under certain obligations to the landlord, including receiving supplies from the brewery or its approved suppliers. Moreover, the system represents vertical forward integration and accounts for many public houses in England and Wales but is less common in Scotland. Hence, as an alternative to tenancy, the brewery-owned public houses may be operated under direct management, when the manager is a salaried employee with no financial stake in the business.

Time Series Analysis

A statistical approach using historical data development of a variable (e.g., past numbers of tourist arrivals at a destination) to forecast its future development. Therefore, various techniques are available to break down a time series into trend, cycle, seasonal, and random elements. This is therefore a widely used approach in travel, tourism, and hospitality, as well as in another forecasting, of value in producing short-term forecasts. However, the forecasts tend to become less accurate and reliable the longer the time scale.

Time Zones

Divisions of the world into zones predicated on the Greenwich meridian, in which local standard time is applied. As the earth makes a consummate rotation on its axis through 360° of longitude every 24 hours, there are 24 time zones. Therefore, the mean solar time in each zone is at its central meridian, generally 15° apart from the next, but authentic zones correspond to political units rather than rigorously following the meridians. Hence, astronomically immense countries with a wide east to west spread are divided into several time zones, e.g., the continental USA into five, Canada into six, and the Russian Federation into 11.

Timesharing

Each of these parties is entitled to inhabit the property for one or more predetermined periods each year while they are renting a furnished and serviced holiday (vacation) home. This is the most common meaning of the term and although a timeshare is said to be owned. Therefore, there are important distinctions between mere timesharing, which gives no more than a right to use the property, and ownership, with such rights as to rent or sell, and variants of both. Also called holiday (vacation) ownership, interval ownership, and multi-ownership, especially in Australia and the USA. It was estimated in 2000 that there were 4 million timeshares owners worldwide, more than 1 million of them in Europe (the second largest number after the USA), of whom more than 300,000 were in the British Isles.

Tip

Term said to have originated in early English coffee-houses as an acronym formed from the first letters of 'To Ensure Promptitude.' This term is synonymous with gratuity.

Toastmaster

Person, often uninformed, employed at a formal banquet for such functions as announcing the names of arriving guests and calling for silence for speakers to speak.

Toll

The tax is imposed on users of certain roads, bridges, canals, and other permanent ways to pay for their construction and/or maintenance.

Toll-Free Number

Callers will be able to make a free, at the expense of the subscriber, call beginning with 1–800 in North America, 0800 in the UK.

Ton (Tonne)

Several different meanings can be attributed to this measure of weight:

- In avoirdupois, 1 long ton (lgt) equals 2,240 lb (20 cwt of 112 lb) or 1.016 metric ton (ton);
- A metric ton (ton) equals 1,000-kilogram(me)s or 2204.62 lb or 0.984 long ton or 1.1 short ton;
- A short ton (American ton) equals 2,000 lb (20 cwt of 100 lb) or 0.907 ton or 0.892 long ton.

Tonnage (Shipping)

There are three chief proportions of the size of a boat in weight terms:

- Deadweight weight estimates the load in lengthy tons (2,240 lb) a boat can lawfully convey, i.e., the complete load of freight, fuel, stores, travelers, and team.
- Gross weight estimates the limit, i.e., with specific exemptions the complete volume of all encased spaces of a vessel (inside the structure or more the deck accessible for freight, stores, travelers, and group), the unit of estimation being a huge load of 100 cubic feet (100 cubic feet of limit is identical to one gross ton).
- Net weight is gotten from the gross weight by deducting spaces for team convenience, driving apparatus and fuel.

Topography

Portrayal or portrayal on a guide of extremely durable/semi-long-lasting regular and man-made/altered surface of an area, i.e., both landforms and other normal elements as well as those delivered by man.

Tornado

A brutal tempest as a rule of brief span, normal in West Africa toward the start and end of the stormy season and in the USA among April and July (Figure 79).

Figure 79. Tornado storm.

Source: Image by Flickr.

Total Quality Management (TQM)

A way to deal with working on the viability and productivity of business including the entire association to guarantee consumer loyalty through zeroing in on client needs and the goals of the association. Therefore, the orderly interaction comprises of deciding the clients' necessities; creating offices and administrations which will meet those prerequisites; working them in congruity with the laid-out norms; observing the consumer loyalty. Hence few aircrafts, Disney, and Holiday Inns have driven the way towards TQM in movement, the travel industry and accommodation.

Tour Operator

An individual or association purchasing individual travel administrations (like transportation and convenience) from their suppliers (like transporters and lodgings) and consolidating them into a bundle of movement. Therefore, the visit is sold with an increase to the public straightforwardly, or through go-betweens. Albeit in some cases depicted as a distributor (visit distributer in USA), a visit administrator is, truth be told, a maker of movement items, whose exercises might be contrasted with those of others essentially gathering item parts, like engine vehicle producers or, to be sure, book distributors.

Tour-Basing Fare

A decreased full circle air or ocean passage accessible to visit administrators for use in comprehensive visit development.

Tourism Accommodation

There is no generally acknowledged meaning of 'the travel industry convenience. Yet it could be viewed as an office that consistently (or sometimes) gives for the time being convenience to sightseers. Therefore, the travel industry convenience is partitioned into two principle gatherings: aggregate the travel industry foundations and private the travel industry convenience [World Tourism Organization (WTO)].

Tourism Activity Index

Proportion of relative change in the travel industry action over the long run, in which participation information at given areas are utilized as a proportion of the travel industry level. Otherwise called the travel industry indicator

[Huan, T. C. furthermore, O'Leary, J. T. (1999) Measuring Tourism Performance, Champaign, IL: Sagamore Publishing].

Tourism Attractiveness Index

Proportion of the travel industry capability of various districts, credited to G.E. Equipping, W.W. Swart and T. Var. The methodology includes requesting a board from specialists to appoint loads to a progression of properties to mirror their general significance for the travel industry advancement and requesting that the specialists assess every area on these qualities [Smith, S. L. J. (1989) Tourism Analysis: A Handbook, London: Longman].

Tourism Balance

Distinction between worldwide the travel industry receipts and global the travel industry consumptions of a country. In like manner, nations might be separated into those with a positive travel industry balance (especially Austria, France, Italy, and Spain) and those with a negative travel industry balance (especially Germany, Japan, the Netherlands, United Kingdom). Moreover, generally viewed as a proportion of the net effect of global the travel industry on a nation's equilibrium of installments, however, it does exclude such impacts as spillages.

Tourism Concentration Index

Proportion of the travel industry fixation as a proportion of evenings spent in a space from specific areas of beginning credited to M. Jensen-Verbeke, which shows the level of reliance of the region or the local centralization of the market [Tourism Management, Vol. 16, No. 1, February 1995].

Tourism Destinations

Nations, locales, towns, and different regions which draw in sightseers, are principal areas of traveler movement, and will quite often represent the majority of vacationers' time and spending. Therefore, they are the principal convergences of vacation destinations, convenience, and other traveler offices and administrations, where the fundamental effects of the travel industry – financial, social, and physical – happen.

Tourism Development Action Plans (TDAPS)

An organization of drives laid out in England by the English Tourist Board (ETB) during the 1980s to foster the travel industry in areas of potential and need. TDAPS covered both country and metropolitan regions and depended on associations between the ETB, nearby specialists, other public offices, and the private area. Moreover, regularly north of three years, to lay out for each situation a neighborhood responsibility, to be maintained and advanced in the more drawn-out term with nearby assets. The principal drive began in Bristol in 1984 was trailed by such urban communities as Bradford, Portsmouth, Carlisle, and Lancaster, by such provincial regions as Exmoor and Kielder Water, and via oceanside hotels like Bridlington and Torbay.

Tourism Expenditure

Characterized for measurable purposes as the absolute utilization consumption made by a guest or for the benefit of a guest for and during his/her excursion and remain at an objective. Therefore, the suggested breakdown contains seven primary classes: bundle travel, bundle occasions and bundle visits; convenience; food and beverages; transport; entertainment, culture, and brandishing exercises; shopping; other [World Tourism Organization (WTO)].

Tourism Expenditure Impacts

Traveler spending has, initial, an immediate impact on the underlying beneficiaries (e.g., inns, eateries, shops) and on variables of creation utilized by them (land, work, capital). Therefore, a circuitous impact is made by progressive rounds of deals, as provisions are bought inside the economy. Moreover, an instigated impact results from expanded shopper use because of immediate and aberrant impacts. Hence, direct effects are some of the times alluded to as essential, aberrant, and instigated impacts are joined as optional. However, the immediate, aberrant, and initiated effects of the travel industry use on a public, provincial or neighborhood economy might be estimated with regards to exchanges or deals, yield, pay, business, and government income.

Tourism Generating Areas

Areas of beginning of sightseers, i.e., the region of their extremely durable home, which address the wellspring of interest and the area of the market, where the significant advertising elements of the travel industry – advancement, visit activity, travel retailing – are based. Therefore, huge convergences of populace in created nations are the primary producing areas of global and homegrown the travel industry.

Tourism Impact

Overall terms, the impact that travelers and the travel industry advancement have on a local area or region. The effect is generally classified into economic, social, and cultural, and environmental.

Tourism Industry

Term to depict firms and foundations giving attractions, offices, and administrations for travelers. Monetary exercises are ordinarily gathered into ventures as indicated by their items. Therefore, as vacationers utilize a scope of attractions, offices, and administrations, they are clients of various businesses as expectedly characterized. Those altogether reliant upon vacationers for their business, like lodgings and visit administrators, are now and again called the travel industry related enterprises. Nevertheless, to the degree to which they supply travelers as opposed to nearby and neighborhood markets, they make up a travel industry, that piece of the economy that has a typical capacity of addressing vacationer needs.

Tourism Intensity

The connection between the quantity of vacationers or travelers' evenings and the quantity of inhabitants of a location region, once in a while communicated as a proportion by partitioning the previous by the last option and portrayed as vacationer force record [Lundberg, D. E. (1974); The Tourist Business, second release, Boston, MA: Cahners].

Tourism Multipliers

Mathematical coefficients which measure the all-out impact (i.e., immediate, roundabout, and prompted) of beginning the travel industry consumption in a space, because of its resulting dissemination in the economy. Various kinds of multiplier measure the impact on business turnover, the degree of result in the economy, complete livelihoods, work, and government income. Therefore, the multiplier values rely upon inclinations to consume and to import: the higher the extent of pay which is spent as opposed to saved and the lower the import content of the travel industry consumption, the bigger the multiplier as well as the other way around.

Tourism Peaking Index

Measure summing up information on fleeting use levels, ascribed to D.J. Stynes, with a base worth 0.00. The more noteworthy the level of focus over a period, the more noteworthy the worth of the file [Smith, S. L. J. (1989), Tourism Analysis: A Handbook, London: Longman].

Tourism Police

Police selected in certain nations explicitly to help and safeguard sightseers. Accessible to answer enquiries and help in crises. Therefore, they additionally shield travelers from double-dealing by nearby brokers (Figure 80).

Figure 80. Mumbai tourism police jeep.

Source: Image by Wikimedia Commons.

Tourism Ratio Index

A proportion of the connection among approaching and active traveler streams in a space, which surveys the net increase or loss of the travel industry nearby, ascribed to M. Jensen-Verbeke [Tourism Management, Vol. 16, No. 1, February 1995].

Tourism Sector

The piece of the economy which has a typical capacity of addressing vacationer needs, comprising of the travel industry related ventures to the degree to which they supply travelers as opposed to nearby and neighborhood markets.

Tourism Transport

For factual purposes, the method for transport alludes to the means utilized by a guest to go from his/her place of regular home to the spots visited [World Tourism Organization (WTO)]. A proposed order comprises of two levels: the principal level (significant gatherings) alludes to the vehicle course (air, stream, land) and the subsequent level (minor gathering) determines each method for transport inside the significant gathering, (for example, planned flights, traveler lines and ships, rail lines).

Tourism Related Avenues

Term now and again utilized for ventures serving travelers straightforwardly and to a more prominent or lesser degree subject to the travel industry for their business. Therefore, regarding the standard industrial classification (SIC) the principal ventures ordinarily included are lodging and cooking administrations, transport, retail circulation, sporting, and social administrations. These compare comprehensively to fundamental classifications of the travel industry use, i.e., convenience, food, and drink; transport; shopping; amusement and recreation.

Notwithstanding, the correspondence is not even close to exact and, in addition, the SIC does not necessarily recognize independently such exercises exceptionally reliant upon the travel industry as visit activities and travel services.

Tourist

For factual purposes, 'a guest whose visit is for somewhere around one evening and whose principal reason for visit might be ordered under one of the accompanying three gatherings:

- Recreation and occasions;
- Business and expert; and
- Other the travel industry purposes' [World Tourism Organization (WTO)].

Tourist Board

A public, provincial, or nearby association differently worried about the turn of events, advancement, and coordination of the travel industry in its space, which might be an administration office, a legal body or a willful relationship of the travel industry interests. In order to be recognized from sectoral associations, for example, exchange affiliations worried about specific businesses or staff affiliations worried about specific occupations occupied with the travel industry.

Tourist Enclave

A pretty much encased and isolated the travel industry objective region, where vacationers are thought and separated from contact with the occupant populace, once in a while by configuration to keep away from struggle.

Tourist Function Index

A rough proportion of the significance of the travel industry in an area, credited to French geographer Pierre Defert, and communicated as a proportion of the quantity of accessible traveler beds and the inhabitant populace. (Likewise the travel industry power [Smith, S. L. J. (1989), Tourism Analysis: A Handbook, London: Longman]).

Tourist Information Center (TIC)

Office offering data about vacationer/guest attractions, offices, and administrations, and some of the time likewise offering such administrations as convenience reservations.

Might be given by government, vacationer sheets or another association and have a neighborhood or more extensive degree. Spasms structure an exceptionally evolved and coordinated network in various nations, including the UK.

Tourist Product

From a tight perspective, what sightseers purchase, e.g., transport or convenience, independently or as a bundle. From a more extensive perspective, a combination of what the vacationer does and of the attractions, offices, and administrations he/she uses to make it conceivable.

According to the travelers' perspective, the all-out item covers the total insight from venturing out from home to return. As particular from an aircraft seat or a lodging as individual items, the all-out traveler item – be it an oceanside occasion (excursion), a touring visit or a meeting trip – is a composite item.

Tourist Tax

Any obligation, duty or assessment gathered by focal or neighborhood government, their offices, or different specialists from guests, in such structures as bed expense, flight or section duty, lodging or room charge, resort duty or guest charge.

Tourist/Visitor Attraction

Components of the vacationer item which draw in guests and decide the decision to visit one spot instead of another. Essential qualifications are between site attractions (e.g., climatic, beautiful, authentic) when the actual spot is the significant actuation for a little while, and occasion attractions (e.g., celebrations, games, exchange fairs). Moreover, when the occasion organized is the bigger component in the vacationer's decision than the site; frequently the site and the occasion together join to decide the travelers' decision. Yet another differentiation is among regular and man-made or fabricated attractions, as among seashores and legacy towns.

Townhouse Hotel

Kind of little, ordinarily exclusive inn offering extravagance convenience in a town place in a reproduced constructing.

Trade Association

A willful non-benefit making collection of free firms in a specific exchange or industry which exists to secure and propel their normal advantages through portrayal and arrangement of administrations to individuals to aid the direct of their organizations. Therefore, chief exchange relationship with an interest travel, the travel industry and neighborliness depend on part enterprises, and incorporate lodging and providing food, transport, visit administrator and travel planner affiliations.

For models see those recorded under Trade relationship in neighborliness industry (UK); Trade relationship of visit administrators and travel planners (UK); Trade relationship of transport administrators (UK).

Trade Mission

A gathering visit by business or potentially government delegates to at least one nation to advance new business for their product(s), administrations, or objections.

Trade Union

A relationship of workers in a specific exchange or industry, or of specific representatives in more than one industry, whose chief capacities incorporate guidelines of relations among them and businesses or managers' affiliations. The three principal types are: the modern association (covering one industry with practically zero-word related qualification); the specialty association (which puts together basically talented workers in a specific occupation, in anything that industry they might be found); the overall association (which remembers for its participation for the most part incompetent and semiskilled representatives in more than one occupation). Hence, chief worker's organizations with an interest in movement, the travel industry and accommodation mirror the three sorts in many nations. For models see those recorded under trade associations in movement, the travel industry and accommodation (UK).

Trade Wind

Wind blowing constantly towards the equator between around 30° north and 30° south scope in the Atlantic and Pacific while being deflected towards the west by the earth's rotational force, i.e., Coriolis force.

Trading Down

Selling at a lower cost, typically joined by a decrease of value or level of administration, to accomplish a higher volume by drawing in additional clients. In this way, e.g., a change to self-administration at lower costs in a current café might be viewed as 'exchanging down,' particularly whenever joined by decrease of decision and additionally accessibility of 'value' dishes.

Trading Up

Selling at a more exorbitant cost, generally joined by worked on quality or level of administration, with the end goal of accomplishing higher overall revenues. In this manner, e.g., an inn repair bringing about arrangement of ensuite offices at a higher levy would address exchanging up.

Traffic Rights

Broadly talking, freedoms haggled by reciprocal air administrations arrangements between states managing global air administrations between them. See: additionally Bermuda Agreement; Chicago Convention; opportunities of the air. More specifically, the Third, Fourth, and Fifth freedoms of the air, as distinct from the First and Second freedoms, known as technical rights.

Trail

In amusement and the travel industry, most normally an assigned course with signposting to direct walkers, cyclists, drivers, or skiers along it. Albeit predominantly created to help guests, trails are additionally progressively utilized as a method for 'making due' guest streams in high thickness regions.

Trailer

That which trail, e.g., trailer train (towed band), trailer interview/question/ study (that joined on as a feature of a bigger meeting/poll/overview).

Trans-Tasman

Term used to portray connections among Australia and New Zealand, nations isolated by the Tasman Sea. E.g., trips between the two nations are alluded to as Trans-Tasman flights; these are worldwide flights however under the arising single flying business sector in Australasia have moved towards homegrown status. (See Australia New Zealand Closer Economic Relations Trade Agreement).

Trans-Siberian Express

The train working across Asia, initially from Moscow to Vladivostok, along the world's longest railroad line, which covers almost 100 levels of longitude, seven-time regions and 5,900 miles (9,500 kilometers).

Transfer

Administration for showing up and leaving travelers to ship them among air terminals and ports, air, ocean, and rail terminals and lodgings, or between transport terminals, given via transporters, inns, or different administrators, normally free between air terminal terminals, as inn civility administration or as a feature of a comprehensive visit.

Transit Hotel

A portrayal now and again utilized for an inn taking special care of short-stay visitors on the way to different objections (Figure 81).

Figure 81. Louis' Tavern transit hotel, Suvarnabhumi airport terminal, Bangkok.

Source: Image by Flickr.

Transit Traveler

In wide terms, an explorer going through an area, which isn't his/her final location. In worldwide travel, normally characterized for specialized reasons as one who doesn't officially enter the country, for example, an air traveler staying in an assigned region of the air terminal or boat traveler not allowed to land; this may yet require not be the definition utilized specifically the travel industry measurements.

For nations with broad land boondocks, for example, Austria, the Czech and Slovak Republics and Switzerland, travel traffic by street frequently addresses a high extent of guests with a huge consumption on the way.

Transport Account

A part of the ebb and flow record of a nation's equilibrium of installments comprised of isolated ocean transport and common avionics accounts.

For instance, in the UK transport account ocean transport covers exchanges of UK administrators with abroad occupants and of abroad administrators with UK inhabitants; common flying covers abroad exchanges of UK carriers and the exchanges of abroad aircrafts with UK inhabitants.

Travel Account

A part of the ongoing record of a nation's equilibrium of installments showing profit from and use on global travel barring worldwide vehicle, i.e., sums spent in the country by occupants of different nations and by the country's inhabitants in the nations visited.

Major getting nations will quite often have a positive equilibrium, major producing nations a negative equilibrium on movement account.

Travel Advisory

Advice, frequently a warming, given by an administration authority (e.g., the Foreign Office in the UK or the State Department in the USA) with respect to head out to a nation or region, in such cases as common agitation or wellbeing danger.

Travel Agent

An individual or association selling travel administrations (like transportation, convenience, and comprehensive visits) in the interest of administrators (like transporters, lodgings, and visit administrators) for a commission. Therefore, most travel planners additionally regularly offer auxiliary types of assistance. For example, getting identifications and visas, secured checks (secured checks), monetary standards and travel protection. Moreover, the chief elements of the travel planner are those of a retailer. In order to give admittance to a head to the market and to give an area to the client to purchase travel administrations.

Travel Compensation Fund (TCF)

A focal asset to which all Australian authorized travel planners are expected to contribute. The asset is utilized to repay explorers in case of travel planner bankruptcy yet it does not cover misfortunes caused by the breakdown of chiefs.

Travel Coupon

A coupon of a transporter's ticket which is gathered from the traveler when the excursion is embraced.

Travel Document

- Passport, visa, and some other report expected to empower an individual to enter or leave a nation or to guarantee the utilization of specific offices and administrations in transit.
- A record gave to stateless people by the specialists of their nation of home to travel abroad, which needs to incorporate a visa for all nations to be placed.

Travel Fair

A show and meeting place for providers, for example, attractions, and convenience suppliers and visit administrators and purchasers like travel planners or the overall population. A differentiation might be drawn appropriately between movement fairs from which the overall population is or alternately is not rejected. Hence, in some both are obliged, normally by conceding them on various days.

Travel Industry

From a restricted perspective, traveler transport transporters and firms and foundations sell their administrations including visit administrators and travel planners. From a more extensive perspective, all organizations and foundations serving the necessities of explorers, additionally portrayed as movement exchange, now, and again utilized equivalently with the travel industry.

Travel Insurance

Contract giving, regarding a premium, for the installment of an amount of cash in case of certain travel-related happenings. A movement insurance contract regularly covers as a base:

- Individual mishap;

- Clinical and related costs;
- Crossing out and abbreviation;
- Travel deserting and delay;
- Possessions and cash;
- Individual responsibility.

Travel Voucher

A generally useful voucher gave by a visit administrator to someone else or association, for example, ground taking care of specialist, to cover prepaid components of a bundle visit. Therefore, the voucher might cover admission to traveler/guest attractions, nearby vehicle, dinners, or different things.

Travel Warrant

Voucher giving composed approval to a transporter to give a pass to its holder. Given under a plan by an association with the transporter, e.g., by a tactical unit for military workforce to get a pass to go by rail.

Traveler

Any individual out traveling between at least two nations or between at least two regions inside his/her nation of regular home. A qualification is made for reasons for the travel industry measurements between two expansive kinds of voyagers: 'guests' and 'different explorers.' Moreover, a wide range of explorers occupied with the travel industry are portrayed as guests. Hence, the term 'guest' addresses the essential idea for the entire arrangement of the travel industry measurements [World Tourism Organization (WTO)].

Traveler's Cheque (Traveler's Check)

A type of check utilized by voyagers, particularly for movement abroad. Given by banks, enormous travel organizations and different foundations. In a few monetary forms and in different divisions, secured registers are typically convertible with monetary standards of the nations where they are encased. Therefore, each check must be endorsed by the buyer at the hour of issue, and countersigned when encased. In case of misfortune or burglary, most giving associations embrace to make a discount to holders, gave basic insurances have been noticed. Furthermore, secured checks are a typical method for installment for movement, the travel industry and friendliness administrations around the world.

Traveler's Diarrhea

Nonexclusive expression for a wide scope of gut diseases, brought about by various creatures, and spread by defiled food and water. It covers most normal travel-related diseases with some level of chance in many regions of the planet, particularly in agricultural nations. Therefore, the principal types of anticipation are severe cleanliness and care with food and water.

Treaty of Amsterdam

Result of the 1996 European Union (EU) between administrative gathering met to consider Treaty revisions, including joining of the Social Chapter into the Treaty, which went into force in 1999.

Treaty of Rome

Understanding endorsed in 1957 by the six establishing nations of the European Economic Community (EEC) making a provincial gathering with the essential points of free development of merchandise, administrations, capital, and individuals between part nations (France, Germany, Italy, Belgium, Netherlands, Luxembourg).

Treeline

The climatic tree line addresses the furthest reaches of tree development under normal circumstances and differs with direction, size of the upland mass and the idea of the environment. It is therefore a lot of lower in the maritime environment of the United Kingdom (UK), arriving at nearly ocean level in the Western Highlands and just 2,800 feet (850 m) in the Cairngorms, more than in the huger size and mainland environment of the Alps. The business tree line is thus a lot of lower since not entirely settled by business contemplations.

Trespass

Unlawful passage to the property of another, significant in a sporting setting corresponding to exercises on rustic land. The lawful circumstance changes impressively from one country to another and the meaning of trespass as an issue relies generally upon how much open land accessible, its area, the power of purpose of private land and the thickness of populace. Therefore, in England and Wales, where maybe 10% of the land is openly claimed/made due, trespass is a common offense, albeit late regulation has recognized a class of disturbed trespass. Moreover, trespass is confounded by the organization of privileges of way across private land, which give just freedoms of entry.

In spite of a broad popular assessment that there ought to be a right of admittance to open country in the uplands, this doesn't exist; access arrangements can be haggled by nearby specialists with private landowners, yet not many as a matter of fact have been made. Moreover, in the USA, where there is significantly more open land for entertainment and no organization of freedoms of way across private land, mentalities to intruders will quite often be a lot harder.

TRINET (Travel Industry Research Information Network)

The travel industry research information network, a global electronic organization facilitated by the School of Travel Industry Management related to the Computing Center of the University of Hawaii. Therefore, the organization works with trade of data among endorsers on research projects, references, meetings, and their procedures, postulations, award recommendations and subsidizing, and the travel industry instruction, using telecom joins being used in colleges around the world.

Trip

There is no commonly acknowledged meaning of an outing, and the term will in general be characterized for specific purposes by reference to such models as distance, length, and reason. E.g., Statistics Canada and Tourism Canada utilize a base distance of 80 km (50 miles), US Bureau of Census and US Travel Data Center 160 km (100 miles). Moreover, in UK studies a stay of at least one evening from home for most purposes will in general be the meaning of a traveler excursion, and trips there and back enduring no less than 3 hours are utilized to characterize road trips and visits.

Trip Index

A factual procedure ascribed to D.G. Pearce and J.M. Elliott, to analyze the degree to which spots visited by vacationers are significant locations or simply visits, and determined by isolating evenings spent at the location by the complete number of evenings spent out traveling (3,100). Therefore, a list worth of 100 demonstrates that the whole excursion was spent at one objective, a worth of zero that no short-term visit was made on the outing [Journal of Travel Research, *32*(1), pp. 6–9].

Trishaw

A vanishing type of three-wheeled bike transportation for conveying individuals and a colorful type of transport for sightseers utilized in Asia.

Tronc

A customary technique for pooling and sharing tips in eateries by which cash got by workers from clients is paid into a typical asset and conveyed at ordinary stretches on a set up premise, regularly utilizing a focuses framework. Therefore, the individual responsible for the asset and its conveyance is known as the truck master.

Tropics

The zone of the world's surface between the Tropic of Cancer and the Tropic of Capricorn, i.e., between 23°30' north (Tropic of Cancer) and south (Tropic of Capricorn) of the equator, wherein the sun is upward at noontime for quite a while during the year. Therefore, the term additionally alludes to the zone outside the central belt with its own unmistakable sorts of environment, vegetation, and different qualities of significance for movement and the travel industry. Moreover, these regions incorporate the absolute most popular worldwide the travel industry objections as, e.g., the main part of the Caribbean, Hawaii, and other Pacific islands.

Trunk Roads

Fundamental streets; in the UK, the public organization of through courses, for which the Secretary of State for Transport is the interstate power. Moreover, the organization comprises of all-purpose trunk streets, which are open for use by all classes of traffic, and extraordinary streets. For example, motorways, which are held for use by indicated classes of traffic as it were. District boards are the thruway experts for public streets which are not trunk streets.

Tuberculosis

A sickness once viewed as essentially killed yet presently expanding around the world, particularly in Eastern Europe, Asia, Africa, Central, and South America. Those remaining over a month in those areas, ought to examine the requirement for vaccination with their PCP, ideally something like two months prior to voyaging.

Tundra

A treeless locale between the tree line and polar ice in Eurasia and northern Canada, with long serious winters and permafrost, where even the mean month to month summer temperature is under 10°C (50°F). Therefore, turning down am afternoon practice of certain lodgings setting up the bed for use by eliminating the comforter and in some cases additionally including different undertakings like closing shades and supplanting utilized restroom material (Figure 82).

Figure 82. Tundra plateau with a small herd of reindeer.

Source: Image by Flickr.

Turning Down

An afternoon practice of certain lodgings setting up the bed for use by eliminating the blanket and now and again additionally including different errands like closing drapes and supplanting utilized restroom cloth.

Turnkey

Term utilized in mix with different words to indicate something finished to the moment that the client just needs to 'turn a key' to make it functional. In this manner, e.g., a turnkey PC framework is set to play out a total arrangement of methodology when it is enacted. Therefore, a turnkey development contract passes on the project worker to see to all subtleties and hand over a functional unit.

Turnpike Road

A street on which obstructions known as expressways are or alternately were raised for the assortments of costs required on clients, consequently a principal street or parkway as of now or previously kept up with by tolls for the purpose of charging the expense for the genuine clients. In North America, likewise called expressway rather than turnpike.

Twinning

A term given to a social relationship framed by city pioneers between two towns or urban communities (alluded to as sister towns or urban areas) in various nations to advance contacts and generosity between their occupants. This gives a premise to bunch visits and different types of collaboration and consequently an improvement to travel and the travel industry.

Typhoid

A contamination brought about by a salmonella infection, which shows itself in fever. It is caught through the utilization of tainted food or water, and the gamble is most elevated in non-industrial nations yet is not bound to them. Despite an antibody is accessible, severe cleanliness and care with food and water are encouraged.

Typology

A plan of order of types gathered by unambiguous models. Consequently, a vacationer typology reflects sorts of sightseers as indicated by such models as inspirations, interests, and styles. Notwithstanding their scholarly premium, typologies are likewise of functional significance in portraying market specialties as a reason for advancement.

U

Ugli

A citrus fruit similar to a grapefruit. Comparatively, a little larger and of uneven shape. Considerably easy to peel.

Ultra-Heat-Treated Milk (UHT Milk)

Milk which has been treated by sanitizing at temperatures above 135°C, and afterward put aseptically ınto holders.

Ultraviolet Lamp

A light which radiates bright beams which tan the skin, assists the skin with creating Vitamin D, and kill microbes.

Ultraviolet Radiation

Short undetectable beams, past the violet finish of the variety range, which structure the tanning and consuming component in daylight.

Umbrella

A round outline covered with fabric which you hold over your head to keep off the downpour.

Unclaimed Baggage

Cases which have been left with somebody and have not been gathered by their proprietors Unclaimed property or Unclaimed stuff will be sold by closeout following a half year.

Underpass

Where one street goes under one more. A turn is avoided at the lights when you accompany regard to the underpass.

Underplate

A level plate on which one more profound dish is put prior to serving, for example, a plate under a soup plate or under an espresso mug and saucer.

Undertow

A solid momentum close to the lower part of the ocean near the shore streaming the other way to the surface flow brought about by the water tossed on the shore streaming back; it addresses a threat to swimmers.

Uniform System of Accounts for Hotels

An arrangement of rules for the planning and show of lodging accounts in a standard structure. First distributed by the Hotel Association of New York City in 1926 and presently in its 10th changed version. The framework is therefore in wide use all through the world and empowers information in light of it to be thought about between inns. Hence, comparable frameworks created in a few nations, including the United Kingdom (UK), have a more restricted use.

Uniformed Staff

Term used to depict by and large inn representatives giving front of-house individual administrations to visitors, including usually a head lobby doorman (ringer skipper), watchmen (chime jumps) and page young men (ringer young men) yet additionally others in certain lodgings.

Unique Selling Proposition (USP)

A publicizing idea ascribed to Rosser Reeves, American promoting marketing specialist, depicting a specific item trademark, which recognizes it from contending items, can be viewed as selective to it, and addresses the fundamental motivation to purchase the item. This might therefore be utilized as a promoting topic, which might be held over a significant stretch or another one might be created now and again.

However, noticeable models in movement, the travel industry and neighborliness have included: 'We put in more effort' (Avis); 'World's number one aircraft' (British Airways); 'Just a single lodging network ensures your room will be correct' (Holiday Inns).

Unitary Authorities

Neighborhood experts in England and Wales which consolidate liabilities somewhere else split between province, precinct, and region committees.

United Kingdom Tourism Survey (UKTS)

A month-to-month test request by private meetings as a feature of an omnibus review to quantify travel for all reasons by UK occupants, charged mutually by the public traveler sheets for England, Scotland, Wales, and Northern Ireland starting around 1989, when it supplanted the month-to-month British tourism survey (BTS). Like its ancestor, it covers travel by grown-ups and going with offspring of at least one evening from home. Moreover, a synopsis of the outcomes is accessible in the Digest of Tourist Statistics, distributed by the British Tourist Authority (BTA) and in The UK Tourist, distributed by the English Tourism Council (ETC).

United Nations Development Decade

Portrayal applied to the 1960s when much consideration of individual legislatures and of global associations was centered around the arrangement of improvement plans, in which the travel industry generally assumed a huge and frequently predominant part.

United States Dollar (US$)

Unit of cash of American Samoa, British Virgin Islands, East Timor, Ecuador, El Salvador, Guam, Marshall Islands, Micronesia, Northern Mariana Islands, Palau, Puerto Rico, Turks, and Caicos Islands, United States of America, United States Pacific Islands, United States Virgin Islands (Figure 83).

Figure 83. United State dollar – currency notes.

Source: Image by Pixabay.

Unleavened Bread

Bread made without utilizing a raising specialist like yeast, made in Mediterranean nations, and in India and Pakistan.

Unlimited Mileage

A stipend with a recruited vehicle, where the driver isn't charged for the quantity of miles covered.

Unmade Room

A lodging which has not been arranged by cleaning, changing the cloth and cleaning, and so forth, since the last visitor left it.

Unofficially

In a manner which has not been endorsed by a division or association. For example, he was told by the Foreign Office informally that it would be better assuming he left the country.

Unofficial Strike

A strike by neighborhood laborers, which has not been endorsed by the principal association.

Unpasteurized Milk

Milk which has not been treated by purification Cheese produced using Unpasteurized milk tastes better.

Unrefined

Not handled to eliminate debasements or undesirable substances.

Unskilled Staff

Representatives who have no specific expertise, thus might do general positions like cleaning, washing dishes, conveying merchandise, and so forth.

Unsocial

To work unsocial hours to work now and again when the vast majority are not working. For example in the evening, around evening time or during public occasions.

Unsweetened

Served, cooked, or produced with no additional sugar or other regular or counterfeit improving specialist.

Update

Data which brings something state-of-the-art. They gave a report on the snow estimate. Action word/p det/to re-examine something with the goal that it is more cutting-edge. They have therefore refreshed their manual for Greece to incorporate current costs.

Upgrade

To change to an unrivaled norm, as in moving a traveler or a lodging visitor or their booking to a predominant seat or convenience.

Upholstery

- Crafted by covering seats or different household items with cushioned seats and covers. He has an upholstery shop, fixing old seats and couches;
- Covers for seats, or cushioned seats and pads the upholstery matches the variety plot in the living room.

Up-Market

More costly and interesting to a more modern segment of the populace the organization has chosen to send off a more up-market form of the item. The new shop is exceptionally up-market and everything is incredibly costly.

Upstairs

Towards or in the upper piece of a structure or vehicle Let's go higher up on the top deck: you can see London much better. A modifier in the upper piece of a structure or vehicle. There is a higher up bar for visitors. Thing the upper piece of a structure or vehicle.

Urban Renewal

The rebuilding, redesign, and improvement of old metropolitan regions, in which relaxation and the travel industry frequently assume a critical part. Additionally portrayed as metropolitan recovery.

Urban Tourism

Outings and visits with an emphasis on town and city objections, otherwise called city the travel industry. A quickly developing type of the travel industry, invigorated by verifiable and social attractions, as well as shopping and occasion attractions, and by business travel, it offers a lot of degree for metropolitan recovery. Moreover, notable instances of city recovery through the travel industry remember Baltimore for the USA, Barcelona in Spain, and Glasgow in Scotland (Figure 84).

Figure 84. Image showing urban tourism.

Source: Image by Pixabay.

Urbanization

The process of growth of urban areas so as to account for an increasing proportion of the population living in them. Usually associated with industrialization, the increase in urban population comes about by migration from rural areas and from natural increases. Further, urbanization is of major relevance to tourism, as urban areas tend to have high holiday (vacation) propensities.

Urinal

- A place where men can go to pass waste liquid from the body. There is a public urinal at the corner of the street;
- A bowl to catch waste liquid passed from the body by men The men's room is very modern, with stainless steel urinals.

Use

The act of taking something and making it serve a purpose, or the way in which something is used to make use of to put to a purpose verb /just/ to put to a purpose Don't use the tap water for drinking. Guests used the fire escape to get out of the building.

Use-By Date

A date stamped on the label of a food product, which is the last date on which the food can be safely eaten.

User Bodies

'Watchdog' organizations representing the interests of particular groups of consumers, especially where a supplier has a monopoly or near-monopoly of a public utility such as transport. Moreover, statutory user bodies in transport in the United Kingdom (UK) include the air transport users committee, airport consultative committees, central transport consultative committee and transport users' consultative committees. Hence, voluntary user bodies have been formed, usually on a local basis by pressure groups and some transport operators.

User-Oriented Resources

Resources devoted to recreation and tourism, which depend for their attraction more on their accessibility than on their quality, are located in proximity to concentrations of population, and cater primarily for local or regional needs rather than attracting nationally. Examples include parks, sites for sports or restaurants.

Usual

Normal, happening, done, or taken often the usual method of cooking onions is to fry them gently in a little oil. Therefore, the drink or food which someone well-known to the waiter or barman has most often (informal) A pint of the usual, please. Will you have your usual, sir?

Utensils

A tool or container for use in the kitchen.

U-Turn

A turn made by a vehicle on a road, so as to face in the opposite direction. She made a U-turn and went back to the hotel. U-turns are not allowed on motorways.

UV Fly-Killer

A device which attracts flies to its UV rays and then kills them by electrocution.

V

Vacancy

- An unfilled spot or room 'opening' void rooms accessible 'no opportunities' the lodging or guesthouse is full;
- A task which is not filled We promoted an opportunity in the neighborhood press. We have yet not been able to fill the opportunity for an accomplished top assistant chef.

Vacancy Rate

The typical number of rooms void in a lodging throughout some undefined time frame, displayed as a level of the complete number of rooms.

Vacant

Unfilled or not consumed We have six rooms empty in the annexed. Empty to leave a spot. Guests are approached to clear their rooms before 12.00. To clear the premises to leave a structure, workplaces, or rooms so they become unfilled.

Vacation

US expression for occasion yet in addition utilized in the British Isles for times of formal suspension of typical action, for example, regulation courts and college excursions. See likewise separate passages under occasion.

Vacation Scape

Term used to portray coordinated plan and advancement for the travel industry, begat by American scholastic C. Gunn in his book of a similar title (distributed 1988 by Van Nostrand Reinhold, New York).

Vaccinate

To inoculate somebody against an infection to utilize an antibody to give an individual inoculation against a particular sickness. She was therefore immunized against smallpox as a youngster.

Vaccination

Vaccination with a substance called antibody, which contains adjusted infection or microbes, to give an individual inoculation against a specific sickness. The antibody might be additionally managed in tablet or fluid structure.

Vaccine

A substance used to immunize or inoculate.

Comment: An immunization contains the starts of the sickness, now, and again alive and at times dead, and this is infused into the patient with the goal that their body will foster invulnerability to the infection. Immunization is for the most part given against cholera, diphtheria, rabies, smallpox, tuberculosis, and typhoid.

Vacuum

A space which is totally unfilled of all matter, including air action word to clean for example a room or floor with a vacuum cleaner (casual).

Vacuum Flask

A bottle with twofold dividers which keeps fluids warm or cold. Likewise called bottle cup.

Vacuum-Packed

Stuffed in a unique plastic pack from which all air has been avoided, and afterward chilled or frozen.

Valet Parking

A help at a lodging or eatery where an individual from staff leaves the visitors' vehicles for them.

Validate

- To verify whether something is right the record was approved by the bank;
- To make something legitimate, the ticket must be stepped by the aircraft to approve it.

Validation

Activity of engraving an aircraft ticket with a stamp to make it substantial for movement. Moreover, the mechanical gadget fitted with the transporter's or travel planner's kick the bucket plate used to approve the ticket is known as the validator.

Valley

A long low region, normally with a waterway at the base, between slopes or mountains Fog structures in the valleys around evening time (Figure 85).

Figure 85. View of Parvati valley.

Source: Image by Wikimedia Commons.

Valuables

Things which are worth very much of cash Guests can store assets with the director for safety's sake.

Value

- Worth to be great incentive for cash to be a reasonable arrangement to get an incentive for the means to get a decent arrangement thing of significant worth things which are worth very much of cash Items of significant worth can be saved in the inn safe short-term.
- An amount displayed as a number action word to consider something as being important as an esteemed client, you are qualified for a 10% markdown.

Value Added… Added Value

A monetary idea signifying the worth added by a firm or industry to the expense of its bits of feedback, giving the worth of its result. subsequently, esteem added rises to the distinction between the absolute deals of the firm or industry and the expense of labor and products purchased in.

Value Added Tax (VAT)

An assessment in view of the worth added at each phase of the creation and conveyance of labor and products. Practically speaking the selling cost is expanded by the level of VAT, which must be represented to the specialists, yet against this might be set any VAT remembered for providers' solicitations. Therefore, a tank is utilized all through the European Union (EU) and structures. The reason for the nations' commitments to the Community spending plan yet the paces of assessment as well as things subject to it keep on shifting between the nations. It is additionally by a wide margin the main expense imposed on guests in EU nations.

Vanilla

An enhancing produced using the seed cases of a tropical plant. This formula utilizes vanilla-seasoned sugar. I need vanilla frozen yogurt with chocolate sauce.

Variable Costs

Creation costs which increment with the amount of the item made or administration gave, for example, unrefined components for dinners, visitor room cleaning and material or wages for extra relaxed staff.

Variety Show

A diversion which incorporates a few unique kinds of entertainers like vocalists, artists, conjurors, ventriloquists, and so forth.

Veal

Meat from a calf. *Comment:* Veal isn't as well known in that frame of mind in other European nations.

Vegan

Somebody who does not eat meat, dairy produce, eggs, or fish and eats just vegetables and organic products.

Vegetable

A plant developed for food, not generally sweet, for example, a potato, carrot, onion, cabbage, cauliflower, pea or bean the primary course is presented with a determination of vegetables.

Vegetable Chef

The gourmet specialist responsible for planning vegetables and pasta. Likewise called gourmet expert entremetier.

Vegetable Oil

An oil that has been separated from a plant or the seeds of a plant, for example, olive oil, sunflower oil, sesame oil, or rapeseed oil.

Vegetarianism

There are five primary classes:

- Generally, veggie lovers avoid utilizing meat, fish, and poultry as nourishment for strict, moral, natural or wellbeing reasons;

- 'Demi-vegans' normally reject red meat from their eating regimen however for the most part not white poultry or fish;
- Lactose veggie lovers eat milk and cheddar yet not eggs or anything which has been delivered from a butchered creature;
- Ovo vegans are like lactose veggie lovers yet eat eggs; and
- Vegans eat no creature food or item.

Veggie Burger

A level cake produced using grains, or vegetables, frequently served similarly as a cheeseburger.

Vehicular

Alluding to vehicles (formal) No vehicular traffic is permitted on the island. No vehicular admittance to the motorway vehicles can't get onto the motorway here.

Vending Machine

A machine from which you can purchase something, like desserts, chocolate, cigarettes, or beverages, by placing coins into a space. Likewise called gaming machine.

Venetian Blind

An ignorant concerning shut out light, made of level pieces of plastic or wood, which can be opened and closed or raised and brought down by pulling a string.

Venison

Meat from a deer.

Comment: Venison is passed on to hang for no less than multi-week prior to eating. It is dull red in variety and has next to no fat.

Ventilation Hood, Ventilation Canopy

A gadget set over a stove or other cooking surface, to eliminate scents and steam.

Ventilator

An initial which permits outside air to come in, or a machine which siphons natural air into a room and concentrates flat air (Figure 86).

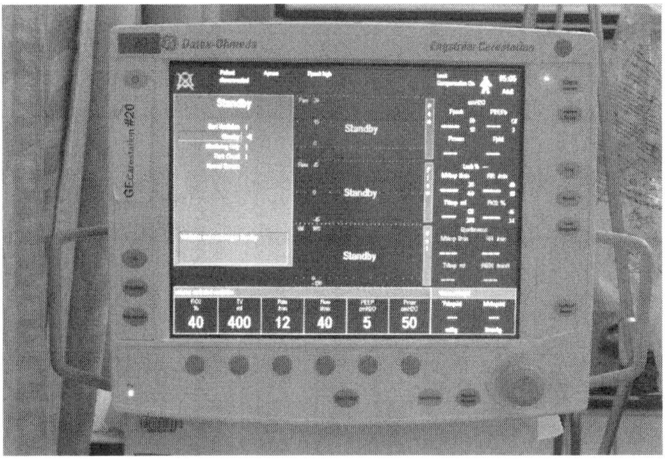

Figure 86. Ventilator control panel.

Source: Image by Flickr.

Vermouth

A sort of solid wine enhanced with spices.

Comment: Vermouth can be either sweet or dry; Italian vermouth is utilized to make martinis.

Vernacular

Language, maxim, expression of one's local spot or nation, i.e., not of unfamiliar beginning or of learned development.

Vertical Integration

The association of organizations that manage various stages in the creation or offer of a similar item, as when a café network assumes control over a wine merchant.

Videotex

Electronic administrations which show communicated data on a video screen. Teletext frameworks, like the BBC Ceefax or the Independent Television Teletext, are communicated frameworks equipped for getting and showing information, and are typically free. Moreover, viewdata frameworks, like the British Telecommunications Prestel or the French Minitel, are phone-based, empower two-way correspondence, and are charged on a for each call premise. Lately, intuitive videotex has upset the dispersion of movement, the travel industry, and neighborliness items, and various administrators work their own frameworks.

Village

A little gathering of houses in the nation, similar to a little town, frequently with a congregation, and generally a few shops.

Vinaigrette

A serving of mixed greens dressing made with oil, vinegar, salt, and different flavorings. Additionally called French dressing.

Vindaloo

An exceptionally hot curry sauce made with coriander, red bean stew, ginger, and different flavors, or a dish cooked in this sauce.

Vinegar

Fluid produced using harsh wine or juice, utilized in preparing and for protecting food red wine vinegar juice vinegar.

Vineyard

A region planted with plants for making wine There are a few grape plantations in southern England. We visited grape plantations along the Moselle and got some wine.

Vintage

- Crafted by gathering grapes to make wine, or the grapes which are gathered;

- Drinks a fine wine made in a specific year This is a generally excellent one of a kind. What classic is it? – it's a 1968. Classic wine, rare port a fine or costly old wine or port which was made in a specific year.

Virus

A minuscule microorganism cell which can create in different cells, and frequently annihilates them Scientists have disconnected another seasonal infection. Remark: Many normal sicknesses. For example, measles or the normal virus are brought about by infections; viral illnesses cannot be treated with anti-infection agents (Figure 87).

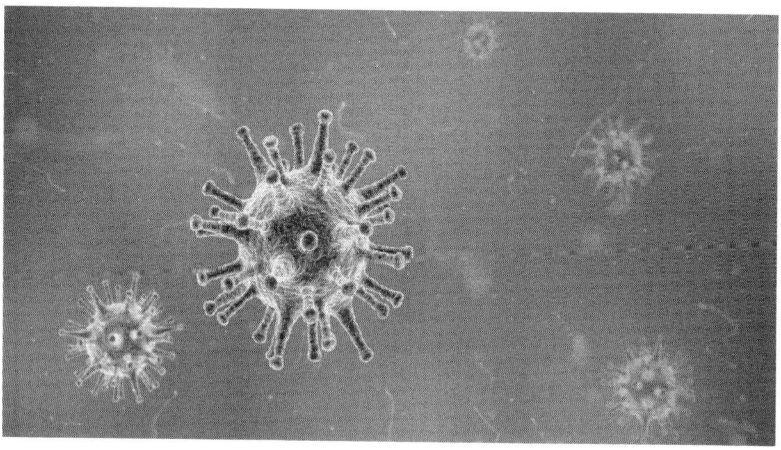

Figure 87. A virus.

Source: Image by Pixabay.

Virtual Reality (VR)

Moderately ongoing improvement in programmatic experiences and human-computer communicating, which endeavors to supplant the client's insight of the actual world with orchestrated three-layered material. This happens through a blend of visual, sound, and dynamic impacts, which cause members to accept that they are really encountering the genuine article.

However, utilized for quite a while in pilot training programs for preparing pilots on the ground, VR innovation apparently has a potential in such regions as the production of virtual amusement parks, as a limited time and deals instrument, and in the formation of counterfeit the travel industry, conceivably in any event, giving in the long run a substitute to travel and the travel industry itself.

Visa

Approval by the public authority, ordinarily entered in the identification, to permit an outsider to enter the country (section visa) or an inhabitant to leave the nation (leave visa), or now and again an occupant voyaging abroad on an outsider identification to reemerge (inhabitant bring visa back).

Passage visas might be of fluctuating legitimacy, single or different section, and characterized by the motivation behind passage, e.g., vacationer, business, travel. Moreover, leave visas are more uncommon and are ordinarily given for a solitary exit (and reemergence if appropriate). Section visas are additionally known in certain nations, especially in Central and South America, as a vacationer card.

Visible

Receipts and installments remembered for the current (as unmistakable from capital) equilibrium of installments account, from merchandise (as particular from administrations).

Visit

A short stay in a spot We will make a short visit to London one week from now. He is on a business visit to Edinburgh. *We had a visit from the VAT investigator*. Action word to go to a spot or to see a person or thing for a brief time frame. For example: he spent seven days in Scotland, visiting historical centers in Edinburgh and Glasgow.

Visitor Center

A structure offering data and administrations to guests in a city or at a verifiable or archeological site, park, or nature hold.

Visitors' Book

A book wherein guests to a historical center or visitors to a lodging record piece of feedback about the spot.

Visiting Friends and Relatives (VFR)

Characterization of fundamental motivation behind trip/visit or movement in like manner use in many nations in fragmenting the market.

Visiting Relatives (VR)

On Australian relocation structures, inbound explorers might choose 'seeing family members (VR)' as their basic role of visit as opposed to visiting companions and family members (VFR), which is the more ordinarily involved classification in different nations. Moreover, the VR classification is the second generally significant (after occasions) because of the enormous foreigner populace with family members abroad.

Visitor

For measurable purposes, any individual venturing out to a spot other than that of his/her standard climate, for under a year and whose fundamental motivation behind visit is other than the activity of a movement compensated from inside the spot visited. Moreover, a differentiation is drawn among worldwide and homegrown guests, and this definition covers two classes of guests: vacationer and immediate guest [World Tourism Organization (WTO)].

Visitor Management

Public and private area frameworks and methodology intended to impact guest conduct at vacationer/guest locales and locations, through such means as limit the executives, understanding, estimating, signposting, and drafting.

Visitor(s) and Convention Bureau

Term of American origination however progressively additionally utilized somewhere else as an assignment for a neighborhood or region vacationer board, e.g., Greater Manchester Visitor and Convention Bureau.

Voodoo

Confidence in and utilization of black magic and so forth ceremonies and strange notions, predominant in the West Indies (WI) and among African Americans (Figure 88).

Figure 88. An altar used in Louisiana Voodoo, on display in the French Quarter of New Orleans.

Source: Image by Wikipedia.

Vitamin

A fundamental substance not orchestrated in the body, but rather found in many food varieties, and required for good wellbeing.

Vitamin A

A nutrient that is dissolvable in fat and can be framed in the body. However, it is the most part found in food, for example, liver, vegetables, eggs, and cod liver oil.

Comment: Lack of Vitamin A influences the body's development and protection from infection and can cause night visual impairment.

Vitamin B Complex

A gathering of nutrients, for example, folic corrosive, riboflavin, and thiamine.

Comment: Lack of nutrients from the B complex can have various outcomes: absence of thiamine causes beriberi; absence of riboflavin influences a youngster's development; absence of pyridoxine causes seizures and regurgitating in children; absence of vitamin B12 causes iron deficiency.

Vitamin C

A nutrient that is dissolvable in water and is found in new organic products, particularly oranges and lemons, crude vegetables, and liver.

Comment: Lack of Vitamin C can cause iron deficiency and scurvy.

Vitamin D

A nutrient which is dissolvable in fat found in margarine, eggs, and fish and furthermore delivered by the skin when presented to daylight.

Comment: Vitamin D aides in the development of bones, and absence of it causes rickets in youngsters.

Vodka

- A solid clear liquor refined from grain or potatoes made initially in Russia and Poland;
- A glass of this liquor Two vodka and tonics, please.

Void

Invalid, which cannot legitimately be utilized, void. Action word to check an archive to demonstrate the way that it can't be utilized legitimately We will void the current ticket and issue another one.

Volcano

A mountain with an opening on the top through which magma, debris, and gas can leave.

Comment: Volcanoes happen along with deficiencies in the world's surface and exist in well-known chains. Some are terminated; however, others eject moderately as often as possible. Some are generally dynamic, in that they discharge sulfurous gases and smoke, without really ejecting.

Volcanic emissions are a significant wellspring of environmental contamination. Volcanoes are famous vacation destinations: the most popular in Europe are Vesuvius, Stromboli, and Etna in Italy and Helga fell in Iceland.

Voltage

A proportion of electrical power.

Comment: The voltage in numerous nations is 110 V (USA, Canada, Japan, and so forth.). In Europe, the far east (FE) and Africa, voltage is generally 220 V or 240 V. Explorers ought to continuously convey a transformer to get electrical apparatuses to work.

Voucher

A piece of paper which is given rather than cash Each explorer has a book of vouchers to be introduced at the front counter in every lodging where the gathering stays.

W

Wadi/Wady

A valley or stream course in the hot desert or semi-bone-dry regions in the Middle East (ME) and North Africa, generally dry however some of the time conveying a stream after a weighty downpour.

Wafer

A slender sweet bread roll is eaten with frozen yogurt, a bowl of vanilla and chocolate frozen yogurt, with a wafer in it.

Waffle

A kind of thick fresh pancake cooked in an iron mold and eaten with syrup. We purchased waffles at the stall in the carnival. Waffles are exceptionally well known in Belgium (Figure 89).

Figure 89. Waffle.

Source: Image by Pixabay.

Wage

Cash paid, for the most part in real money every week or month, to a worker for a work done. She is acquiring a decent compensation or great wages in the bar.

Wage Scale

A rundown of wages, showing different paces of pay for various positions in a similar organization (*Note:* The plural wages is more regular while alluding to the cash procured, however, wage is utilized before different things).

Wagon-Lits

Resting mentors on trains, additionally name of the organization working these, on the Continent of Europe; the mentors include room convenience for a couple of individuals. See additionally couchette; parlor vehicle; Pullman; roomette.

Wait

- To remain where you are, and do nothing until something occurs or somebody comes, we provided our request thirty minutes prior, however, are as yet sitting tight for the principal course. On the off chance that you stand by here, I expect a taxi will show up soon.

- To lookout for somebody CATERING to serve food to somebody at the table.

Waiting List/Waitlist

A rundown of travelers wishing to join a flight, cruising, or visit, which is completely reserved; some of the time additionally utilized by lodgings when full. Those on the rundown may then be obliged if there should arise an occurrence of scratch-offs or flake-outs.

Wanderlust

As indicated by financial specialist H.P. Dark, inspiration to delight travel; the term portrays the craving to trade the known for the obscure, to leave things recognizable and to take a quick trip and see better places, individuals, and societies, i.e., to be presented to what is unique in relation to that which is realistic at home. Moreover, an insatiable craving for novelty or adventure is much of the time-bound to be fulfilled worldwide as opposed to homegrown the travel industry and to call for offices outfitted to short instead of a longer stay. Therefore, the environment is of auxiliary significance and more than one nation or more than one spot is frequently visited.

Warsaw Convention

A peaceful accord was made in 1929 and, as corrected along these lines, restricting the responsibility of aircraft for loss of or harm to stuff (baggage) and injury to or demise of travelers on most worldwide flights (counting homegrown segments of global flights). Because of the Convention, aircraft ordinarily acknowledge responsibility for mishaps up to draw certain lines and petitioners do not need to demonstrate carelessness.

Waste Management

Frameworks and cycles are utilized by organizations and different associations to decrease squander for cost or ecological reasons. Such methodologies frequently rotate round the '3Rs' – reduce, re-use, recycle – and are center parts of cost control and ecological projects in movement, the travel industry, and neighborliness activities.

Water Management

Frameworks and cycles utilized by organizations and different associations to decrease their utilization of water assets for cost or ecological reasons. Such methodologies are center parts of cost control and ecological projects in movement, the travel industry and neighborliness activities.

Water Park

An entertainment region giving water sports, other water-based exercises, and such visual attractions as cascades, for the most part for the overall population and on installment of a confirmation charge. The Cotswold Water Park is a new model in England (Figure 90).

Figure 90. Water park.

Source: Image by Pixabay.

Water Resources Depletion and Pollution

In numerous nations new water assets are as of now scant and are by and large additionally exhausted by high horticultural, modern, and homegrown utilization and by contamination. Most water contamination is brought about by horticultural, modern, and homegrown waste, including insufficient sewage arrangement, with genuine consequences for human wellbeing. Travel, the travel industry and neighborliness exercises are enormous purchasers of water and huge generators of waste; they are therefore straightforwardly impacted by water consumption and contamination in numerous areas, particularly in non-industrial nations.

Water Skiing

A well-known type of dynamic entertainment on lakes and beachfront waters in which the member, mounted on skis, is pulled by a powerboat, expected to produce a speed of around 25 mph (49 kph), adequate to keep up with the skier on a superficial level. It is sometimes a wellspring of contention with different types of entertainment on/in the water, including swimmers, fishermen, and those cruising yachts, and on the shore. Moreover, on openly possessed or controlled water, severe guidelines frequently apply administering where and while water skiing might happen.

Water Sports Holidays (Vacations)

A particular movement occasion (excursion) fragment, including such exercises as dinghy and yacht cruising, plunging, and windsurfing however barring ocean side occasions (get-aways). It is therefore as yet a little extent of all occasions (excursions) – around 1,000,000 were taken in Great Britain (GB) every year during the 1990s. However assessed to become quicker than all occasions (get-aways).

Wayports

Air terminals developed in remote, inadequately populated regions, with practically zero beginning and objective traffic, and committed pretty much solely to taking care of move traffic, with the end goal of limiting ecological issues. Such air terminals might be additionally evolved from military landing strips due for conclusion or from common air terminals whose job has decreased.

Wesak

Significant strict occasion, additionally called Buddha Day, celebrated in Thailand and other Buddhist nations in May every year.

West Country

The area of England ordinarily seen to include the districts of Somerset, Devon, and Cornwall, albeit South West Tourism, one of ten English Regional Tourist Boards (RTBs), covers additionally parts of Dorset, Gloucestershire, Wiltshire, and the Isles of Scilly. This is therefore the most well-known occasion (excursion) district for British inhabitants, which draws in every year a fourth of all occasions of four evenings or more taken in Great Britain (GB) by British occupants, with South of England, Wales, and Scotland being the following most famous objections.

West Indies (WI)

Aggregate expression for the islands of the Caribbean Sea among North and South America.

Wet Rent

The lease payable by the inhabitant to the property manager in the stockpile value, a course of action utilized, e.g., in the tied house framework, when the occupant pays a low fixed lease (additionally named dry lease) for his premises, generally much underneath the market rate, and is charged a more exorbitant cost by the bottling works for the provided lager than is charged to the 'streamlined commerce.' This distinction in cost is named wet lease, rising or falling with the volume of exchange done, expected to change the inhabitant's outgoings to the vacillations of turnover of the public house.

Wetlands

A characteristic or counterfeit scene where the dirt is waterlogged or the land is once in a while, intermittently or for all time covered by new or saltwater. Various kinds of wetlands are portrayed as lowland, peat, bog, or marsh (Figure 91).

Figure 91. Freshwater wetland, Hunter wetlands National Park, Newcastle.

Source: Image by Flickr.

Whale Watching

An arising vacationer movement of developing prevalence in Australia and New Zealand. In Australia, the term covers the perception of whales from boats (as in Queensland) and from shore areas (like the shoreline of Victoria).

Wharf

A stone or wooden construction close by water adjacent to which boats are secured to set out and additionally land travelers or to stack or potentially dump freight.

White Man's Grave

A conversational term previously applied to West Africa, where before the presentation of immunization, prophylactic medications, bother control and further developed sterilization, the blistering sticky environment was especially undesirable for Europeans, bringing about far and wide infection and high mortality. Hence, these days some of the time actually utilized for regions with correspondingly unwelcoming environment.

White-Collar

An American expression used to portray non-manual specialists, specifically administrative and secretarial representatives, presently additionally generally utilized somewhere else.

Wholesaler

Middle person who for the most part purchases merchandise from providers for resale in little amounts to retailers and others. Moreover, in the travel industry the term is some of the time utilized as an equivalent word for visit administrator.

Wind Chill

The impact of wind at low temperatures on concealed dry human skin in making the successful temperature significantly lower. Frequently communicated as a temperature remainder: along these lines a temperature of 20°F (–7°C) and a breeze speed of 45 mph (72 kph) delivers a temperature of –20°F (–30°C). Nevertheless, of extraordinary importance for those climbing, skiing, and journeying at low temperatures, who should guarantee that their apparel is satisfactory for the successful temperatures.

Window-Dressing

Any training that endeavors to cause what is happening to appear significantly more appealing than it truly is. In business utilized especially by bookkeepers and advertising specialists.

Windward

A nautical term indicating the bearing from which the breeze is blowing, i.e., the most un-protected side of a boat.

Winter Sun

Occasions (excursions) intended to exploit the travel industry offices outside the fundamental summer season in regions which have a satisfactory winter environment, like the Mediterranean shorelines of Spain and the Canaries in Europe and Florida in North America, especially appealing to the individuals who have resigned. Furthermore, given variable expenses are covered and some commitment is made to fixed costs, such use is probably going to be advantageous according to the hoteliers' point of view as an option in contrast to cutting off premises during the season months.

Word-of-Mouth Advertising

Publicizing conveyed by a fulfilled client to a family member, companion, or associate, as a forthcoming client for a similar item. Therefore, considered as the most compelling wellspring of data about purchaser items, including travel, the travel industry and neighborliness items.

Work-Study

'The methodical assessment of exercises to work on the compelling utilization of human and other material assets.' Also known as association and strategies (O&M). It comprises of strategy study, characterized as 'the deliberate recording and basic assessment of approaches to getting things done to make enhancements' and work estimation, characterized as 'the use of procedures intended to layout the ideal opportunity for a certified specialist to complete an undertaking at a characterized pace of working' [British Standard Glossary BS 3138 (1979)].

Working Time Regulations

In the UK, long stretches of work and related states of business are ordinarily haggled among managers and representatives on the whole or independently. Be that as it may, European Union (EU) regulations which came into force in the UK in 1998 (prior in other EU nations), which apply to full-time, part-time and impermanent specialists, make the accompanying arrangements (dependent upon exception for laborers in certain areas, including transport, albeit individual specialists can decide to work longer):

- A most extreme typical working seven day stretch of 48 hours;
- At least a month's yearly paid leave;
- Least day to day and week after week rest periods;
- Explicit arrangements for young adult specialists in regard of the above privileges and qualifications.

World Heritage Sites

Regular and social preservation destinations of extraordinary interest assigned by the United Nations Educational, Scientific, and Cultural Organization (UNESCO). Slome models remain: Hadrian's Wall for England, the islands of St Kilda in Scotland, the palaces and town dividers of King Edward I in North Wales, and the Giant's Causeway in Northern Ireland.

World Travel Market (WTM)

A yearly worldwide travel and the travel industry exchange fair held in London in November every year beginning around 1980.

World Span

World span Travel Agency Information Services, US PC reservation framework (CRS) shaped as a joint endeavor between DATAS II and PARS frameworks, which it supplanted in 1990. Delta Airlines, Northwest Airlines and Trans World Airlines (TWA) are head investors, with a little stake by Abacus.

X

Xenophobia

Disdain, abhorrence or feeling of dread toward outsiders or outsiders, or of abnormal or unfamiliar spots.

Xplorer

Australian rail administration connecting Sydney and Canberra and furthermore Sydney and a few communities in New South Wales.

XPT Express

Australian rail administration connecting Sydney and Melbourne, Sydney, and Brisbane, as well as a few New South Wales communities.

Y

Yacht

- A cruising boat, or a boat utilized for delight and game;
- An enormous extravagant boat with an engine. She spent her vacation on a yacht in the Mediterranean (Figure 92).

Figure 92. Yacht in Cape Greco, Cyprus.

Source: Image by Wikimedia Commons.

Yard (Yd)

A proportion of length utilized in English-talking nations is equivalent to 0.9144 meters. A yard is partitioned into 3 feet (ft.) or 36 inches (in.). In proportions of the region, 1 square yard approaches 9 square feet (0.836 square meters) or 1,296 square inches. In proportions of volume 1 cubic yard (cu. yd = 0.765 cubic meter) approaches 27 cubic feet (cu. ft).

Year

A time of the year We are returning to the lodging where we remained the year before. One year from now they are going to Corsica.

Year-End

At the finish of the monetary year, when an organization's records are arranged the records division has begun work on the year-end accounts.

Yeast Extract

A thick tacky earthy colored food acquired from yeast and eaten as a spread or utilized in cooking.

Yellow Fever

Viral contamination causing fever, regurgitating, and liver disappointment; can be deadly. It is caught by the nibble of a contaminated mosquito and happens in pieces of Africa and South America. Immunization is a compelling counteraction for a time a decade. A few nations require immunization authentication for sections.

Yellow Pages

Business phone registry publicizing different items, distributed in many nations around the world.

Yield

- Return based on an interest in conditions of pay corresponding to the ongoing capital worth of the speculation.
- Profit or income inferable from an item, ordinarily communicated as normal benefit or income procured per unit of result, e.g., traveler kilometer in transport or room/night inconvenience.

Yield Management

The idea and procedures worried about the amplification of benefit or income and the improvement of an ideal business blend to those finishes. As of late, yield the executives has gotten expanding consideration and application in transport, particularly in aircraft activities, and in neighborliness the board.

Yorkshire Pudding

A combination of eggs, flour, and milk, cooked in the broiler, the conventional backup to broil hamburger.

Comment: Yorkshire pudding might be cooked on the stove in one enormous dish or in a few little ones.

It used to be cooked in the broiler in a container under the simmering hamburger, with the goal that the juices of the meat trickled onto the pudding. The pudding is generally presented with meat as a component of the fundamental course.

Youth

The European Union (EU) program, inside the general point of permitting youngsters to procure information and abilities and to practice dependable citizenship, incorporates trades between gatherings, individual versatility, youth drives, joint activities, and related help measures. The program is available to 31 nations (15EU, 3EFTA/EEA, 13 EU up-and-comer nations, including Malta and Turkey).

Youth Hostel

A structure where youthful explorers can remain the night inexpensively. If you can track down places at youth lodgings, that is the least expensive approach to voyaging.

Youth Tourism

A fragment of the travel industry market is differently seen to embrace those matured between 15–18 and 25–29 years. Anything definition is embraced; accessible measurements recommend that adolescent the travel industry has been growing quicker than the travel industry in all.

Yukon Standard Time

A Canadian time region in light of the norm of the 135th meridian. The time approaches GMT-8.

Yuppie

The term got from youthful metropolitan experts and indicates an aggressive and effective person in their 30s with an upwardly versatile way of life.

YWCA

An overall government assistance association that runs lodgings giving reasonable convenience to youngsters. Full structure Young Women's Christian Association.

Z

Zabaglione

A pastry made of egg yolks, sugar, and Marsala wine beat over high temp water until pale and frothy and served hot with wiped finger bread rolls (Figure 93).

Figure 93. Zabaglione with cookies.

Source: Image by Wikimedia Commons.

'Z' Bed

A collapsing and versatile bed; when collapsed, it very well may be continued on castors and stand as a household item with a headboard giving a level surface.

Zest

The slender external skin of the strip of a citrus natural product is cut, scratched, or ground to yield a sharp fruity seasoning for food varieties and beverages.

Zionism

The development brought about the foundation of a Jewish state in Palestine.

Zip Code

Shortened form for Zone Improvement Plan code laid out by the US Postal Service to distinguish mathematically the objective of mail for arranging and dispatch.

Zone

By and large, a pretty much-characterized area of land, typically recognized by others by a specific quality or condition demonstrated by a characterizing word, e.g., building zone, duty zone, calm zone. In land-use arranging a region assigned (drafted) for a specific reason, e.g., industry, lodging, entertainment. Subsequently, drafting, i.e., assigning regions for specific purposes.

Zoning Regulations, Zoning Ordinances

Neighborhood standing rules which control the kinds of building and land use in a town.

Zoo

Conversational shortened form for zoological nursery. A site where wild creatures are saved for the general population to see.

Zoological Garden

Where wild creatures are saved so that the general population might be able to see them.

Zwieback

US bread, etc., is a sort of roll made of bread prepared hard.

BIBLIOGRAPHY

1. Getz, D., Abson, E., Antchak, V., Chappelet, J., Goldblatt, J., Jutbring, H., O'Toole, W., Page, S.J., Salama, M., Stadler, R. & Ziakas, V (2021). *Dictionary of Event Studies, Event Management and Event Tourism.* [online] Available at: https://www.goodfellowpublishers.com/academic-publishing.php?content=doi&doi=10.23912/9781911635796-4695 (accessed on 11 May 2022).
2. A. J. B., (1981). The dictionary of tourism. *International Journal of Tourism Management, 2*(4), 275. [online] Available at: https://www.sciencedirect.com/science/article/abs/pii/0143251681900360?via%3Dihub (accessed on 11 May 2022).
3. Biggs, D., (2010). The tourism society's dictionary for the tourism industry. *Tourism Management, 31*(4), 556 [online] Available at: https://linkinghub.elsevier.com/retrieve/pii/S0261517709001228 (accessed on 11 May 2022).
4. Božinovski, B., (2021). All-inclusive LSP dictionaries and the Slovene–English dictionary of tourism. *Language and Tourism,* (pp. 3–32). [online] Available at: https://press.um.si/index.php/ump/catalog/book/635 (accessed on 11 May 2022).
5. Cao, Q., Sarker, M., Zhang, D., Sun, J., Xiong, T., & Ding, J., (2022). Tourism competitiveness evaluation: Evidence from mountain tourism in China. *Frontiers in Psychology* (p. 13). [online] Available at: https://www.frontiersin.org/articles/10.3389/fpsyg.2022.809314/full (accessed on 11 May 2022).
6. Cho, K., (2022). Tourism research paradigms and construction of tourism knowledge. *Journal of Tourism Studies, 34*(1), 57–80. [online] Available at: https://www.dbpia.co.kr/journal/articleDetail?nodeId=N

ODE11038756&language=ko_KR&hasTopBanner=true (accessed on 11 May 2022).

7. Costantini, F., Sidraschi, D., & Zuin, F., (2021). Minority languages as a resource for tourism promotion on the web: the case of some minority communities in Italy. *Language and Tourism* (pp. 33–50). [online] Available at: https://press.um.si/index.php/ump/catalog/book/635 (accessed on 11 May 2022).
8. De-Miguel-Molina, M., Catalá-Pérez, D., de-Miguel-Molina, B., & Santamarina-Campos, V., (2022). The missing link between dark tourism and tourism management. *Tourism Risk* (pp. 221–237). [online] Available at: https://www.emerald.com/insight/content/doi/10.1108/978-1-80117-708-520221015/full/html (accessed on 11 May 2022).
9. Fafurida, F., Daerobi, A., & Riyanto, G., (2022). Implementation model of community-based tourism on rural tourism. *International Journal of Sustainable Development and Planning, 17*(2), 507–512. [online] Available at: https://www.iieta.org/journals/ijsdp/paper/10.18280/ijsdp.170215 (accessed on 11 May 2022).
10. Fauzi, A., (2021). The students' responses on integrating online corpus and online dictionary to expand English vocabulary for tourism. *UAD TEFL International Conference, 2*, 186. [online] Available at: http://seminar.uad.ac.id/index.php/utic/article/view/5756 (accessed on 11 May 2022).
11. Gee, C., (1986). The dictionary of tourism. *Annals of Tourism Research, 13*(4), 672–674. [online] Available at: https://www.sciencedirect.com/science/article/abs/pii/0160738386900150?via%3Dihub (accessed on 11 May 2022).
12. Gugić, J., (2021). 44*Adjective-noun collocations in tourist advertising brochures about Istria: A corpus-based translation study. *Language and Tourism,* (pp. 51–68). [online] Available at: https://press.um.si/index.php/ump/catalog/book/635 (accessed on 11 May 2022).
13. Hillebrand, B., (2022). An ecosystem perspective on tourism: The implications for tourism organizations. *International Journal of Tourism Research*. [online] Available at: https://onlinelibrary.wiley.com/doi/10.1002/jtr.2518 (accessed on 11 May 2022).

14. Jang, K., Sakamoto, K., & Funck, C., (2022). Dark tourism as educational tourism: The case of 'hope tourism' in Fukushima, Japan. *Dark Tourism Studies* (pp. 104–115). [online] Available at: https://www.researchgate.net/publication/357758665_Dark_tourism_as_educational_tourism_the_case_of_%27hope_tourism%27_in_Fukushima_Japan (accessed on 11 May 2022).
15. Khan, S., Nabi, A., & Hafeez, A., (2016). Factors affecting tourism in northern areas of Pakistan. *Journal of Economic Info, 3*(4), 1–6. [online] Available at: https://readersinsight.net/index.php/JEI/article/view/90 (accessed on 11 May 2022).
16. King, B., (1994). Dictionary of travel, tourism, and hospitality. *Annals of Tourism Research, 21*(4), 870–872. [online] Available at: https://www.sciencedirect.com/science/article/abs/pii/0160738394901031?via%3Dihub (accessed on 11 May 2022).
17. Korstanje, M., & George, B., (2022). Tourism, conflict, and conflictivity: Is tourism part of the problem or the solution? *The Nature and Future of Tourism* (pp. 81–96). [online] Available at: https://www.taylorfrancis.com/chapters/mono/10.1201/9781003277507-5/tourism-conflict-conflictivity-tourism-part-problem-solution-maximiliano-korstanje-babu-george (accessed on 11 May 2022).
18. Korstanje, M., (2014). Dictionary of tourism. *Annals of Tourism Research, 45*, 186–188. [online] Available at: https://www.sciencedirect.com/science/article/abs/pii/S0160738313001679?via%3Dihub (accessed on 11 May 2022).
19. Kostanjšek, B. S., (2022). *Art for Tourism.* [online] Available at: https://omp.zrc-sazu.si/zalozba/catalog/book/1999 (accessed on 11 May 2022).
20. Kučiš, V., & Jazbec, M., (2021). Slovenian-German online menus as a treasure trove for translation errors and misunderstandings. *Language and Tourism* (pp. 69–86). [online] Available at: https://press.um.si/index.php/ump/catalog/book/635 (accessed on 11 May 2022).
21. Laskova, V., (2021). Virtual tourism as part of the Russian language classroom. *Language and Tourism* (pp. 109–130). [online] Available at: https://press.um.si/index.php/ump/catalog/book/635 (accessed on 11 May 2022).
22. Lee, Y., & Shin, H., (2022). A study on untact tourism and rural tourism revitalization plan. *International Journal of Tourism Management and*

Sciences, 37(1), 45–55. [online] Available at: https://kiss.kstudy.com/thesis/thesis-view.asp?key=3937615 (accessed on 11 May 2022).
23. Li, M., (2021). Research on extraction of useful tourism online reviews based on multimodal feature fusion. *ACM Transactions on Asian and Low-Resource Language Information Processing, 20*(5), 1–16. [online] Available at: https://dl.acm.org/doi/10.1145/3453694 (accessed on 11 May 2022).
24. Marlinda, A., Al-Fadhat, F., Cipto, B., & Jubba, H., (2022). Halal tourism as a strategic option for South Korean tourism. *Journal of Islamic Marketing*. [online] Available at: https://www.emerald.com/insight/content/doi/10.1108/JIMA-03-2021-0074/full/html (accessed on 11 May 2022).
25. Medlik, S., (2012). *Dictionary of Travel, Tourism and Hospitality*. [online] Available at: https://www.taylorfrancis.com/books/mono/10.4324/9780080521862/dictionary-travel-tourism-hospitality-medlik (accessed on 11 May 2022).
26. Perspectives on Asian Tourism, (2022). *Asian Tourism Sustainability*. [online] Available at: https://link.springer.com/book/10.1007/978-981-16-5264-6 (accessed on 11 May 2022).
27. Potočnik, T. J., (2021). Teaching writing skills in English for tourism by employing travel writing. *Language and Tourism* (pp. 177–196). [online] Available at: https://press.um.si/index.php/ump/catalog/book/635 (accessed on 11 May 2022).
28. Prayudi, M., Sujana, E., Dewi, N., & Wiguna, I., (2019). The research and development study of tourism accounting dictionary. *Proceedings of the International Conference on Tourism, Economics, Accounting, Management, and Social Science (TEAMS 19)*, [online] Available at: https://www.atlantis-press.com/proceedings/teams-19/125924371 (accessed on 11 May 2022).
29. Reindl, D., (2021). Historical names in Slovenian tourism texts. *Language and Tourism* (pp. 155–176). [online] Available at: https://press.um.si/index.php/ump/catalog/book/635 (accessed on 11 May 2022).
30. Roberts, M., (2022). Secondary school students' views of tourism education and tourism careers. *Journal of Hospitality & Amp; Tourism Education* (pp. 1–12) [online] Available at: https://www.tandfonline.com/doi/abs/10.1080/10963758.2022.2056471?journalCode=uhat20 (accessed on 11 May 2022).

31. Sánchez-Cañizares, S., & Cabeza-Ramírez, L., (2022). Restaurants and tourism. *Encyclopedia of Tourism* (pp. 1–3). [online] Available at: https://link.springer.com/referenceworkentry/10.1007/978-3-319-01669-6_162-2 (accessed on 11 May 2022).
32. Sari, D., (2022). Slow tourism possibilities in Cimande tourism village–west java. *E-Journal of Tourism* (p. 31). [online] Available at: https://ojs.unud.ac.id/index.php/eot/article/view/82895 (accessed on 11 May 2022).
33. Smith, W., (2009). The tourism society's dictionary for the tourism industry. *Annals of Tourism Research* (3rd edn., Vol. 36, No. 3, pp. 544, 545). [online] Available at: https://www.sciencedirect.com/science/article/abs/pii/S0160738309000516?via%3Dihub (accessed on 11 May 2022).
34. Uyar, A., Kuzey, C., Koseoglu, M., & Karaman, A., (2022). Travel and tourism competitiveness index and the tourism sector development. *Tourism Economics* [online]. Available at: https://journals.sagepub.com/doi/10.1177/13548166221080357 (accessed on 11 May 2022).
35. Van, D. B. J., (2022). *A Research Agenda for Urban Tourism*. [online] Available at: https://www.elgaronline.com/view/edcoll/9781789907391/9781789907391.xml (accessed on 11 May 2022).
36. Vanhove, N., (2022). Measuring tourism. *The Economics of Tourism Destinations* (pp. 24–63). [online] Available at: https://www.taylorfrancis.com/chapters/mono/10.4324/9781003258186-2/measuring-tourism-norbert-vanhove (accessed on 11 May 2022).
37. Vanhove, N., (2022). Tourism supply. *The Economics of Tourism Destinations* (pp. 145–190). [online] Available at: https://www.taylorfrancis.com/chapters/mono/10.4324/9781003258186-6/tourism-supply-norbert-vanhove (accessed on 11 May 2022).
38. Xiao, H., (2007). A dictionary of travel and tourism terminology. *Annals of Tourism Research, 34*(4), 1095–1097. [online] Available at: https://www.sciencedirect.com/science/article/abs/pii/S0160738307000771?via%3Dihub (accessed on 11 May 2022).
39. Zaitul, Z., Novianti, N., & Ilona, D., (2022). Tourism governance: Village-based tourism criteria in Indonesia. *GeoJournal of Tourism and Geosites, 41*(2), 415–421 [online] Available at: https://gtg.webhost.

uoradea.ro/PDF/GTG-2-2022/gtg.41211-845.pdf (accessed on 11 May 2022).

40. Zavarika, G., (2022). Military tourism as a peculiarity of tourism development in a post-conflict territory. *Journal of Geology, Geography and Geoecology, 31*(1), 186–193. [online] Available at: https://geology-dnu.dp.ua/index.php/GG/article/view/899 (accessed on 11 May 2022).

INDEX

Y